From Biotech to Hollywood

From Biotech to Hollywood

NEW CAREER OPPORTUNITIES IN TECHNOLOGY

DEBRA BENTON

Prentice Hall Press

Prentice Hall Press
A member of Penguin Putnam Inc.
375 Hudson Street
New York, N.Y. 10014
www.penguinputnam.com

Library of Congress Cataloging-in-Publication Data

Benton, D. A. (Debra A.)
From biotech to Hollywood : new career opportunities in technology / Debra Benton.
p. cm.
ISBN 0-7352-0265-6
1. Engineering—Vocational guidance. 2. Technology—Vocational guidance. I. Title.

TA157 .B45 2002
602.3—dc21 2002070017

Printed in the United States of America

10 9 8 7 6 5 4 3 2 1

CONTENTS

ACKNOWLEDGMENTS

Thanks to Fred, Teresa, and Rodney, who have made all the good things in my life possible.

Thanks to my agent, Mike Cohn.

Thanks to Abigail Eslick, Amy Zach Williams, J. M. Jones, Paul Barton, Garrett Credi, Bryce and Rae Land, John Robert Fitzhenry, Melinda Hicks, Killiam Williams, and Dr. Kelvin Kesler.

Please feel free to direct your comments and questions to me.

Debra Benton
Benton Management Resources, Inc.
2221 N. Lake Street
Fort Collins, CO 80521
970-416-3882
Debra@TopSpeaker.com

FOREWORD

Today, the pace of technological advancement is driving the pace of our lives. Bubble or burst bubble, there is still incredible intensity in juggling business and personal time commitments. These continual changes demand constant reprioritization as we try to incorporate more activities into each minute of our day. Whether you are at the beginning of your career, considering a new one, or have had a recent "ah-hah" recognizing the need for career planning, this book is well worth putting aside the Blackberry and turning off the cell phone to invest some of that most precious of assets, your time.

While changes in technology drive changes in how we live our lives, they will also create all kinds of career opportunities. And, these choices will not just be within the technology sector. The ways in which we manage information and knowledge, communicate, and deepen relationships have been transformed, and as a result, so have the ways in which we conduct business across every sector. The fact is that all of us work in technology; without vigilant attention to its advantages and advances, no business or businessperson will realize their potential. This point is one of the key pieces of advice in this book. So whether it's health care, media, financial services, or any industry, you can leverage your passions and be part of this new age.

That fact makes this volume a broad reference book, more than just simply a guidebook on how to get a job in the technology sector. It offers help on how to choose a direction, get practical advice on the process of landing that job you want, or find a career in an organization that is a fit for you. It provides frank and solid direction on every step of the process, as well as insight about the most important aspects of what it takes to be successful in business today and to have a fulfilling career. It drives home the importance of having the courage to learn and to welcome change. The book emphasizes that proactivity, flexibility, adaptability, and an innovative spirit are the personal qualities that will capture the attention of today's employers. The author also appropriately recognizes the value of the old economy's emphasis on general business savvy and interpersonal skills that are so fundamental to business success. The ability to build relationship equity through positive, honest, open, and ethical communication is critical to positioning yourself for a new role and to transitioning into a new company environment. In the end, self-knowledge is the most valuable tool in molding and building a path to leverage your potential toward an outcome consistent with your dreams and values.

I am delighted to have encountered Debra Benton and the work she has done to compile this practical and informative volume on career advice. After eighteen years in executive search, I can say that I have found a book that precisely reflects my own experiences and the recommendations that I would pass on to the great talent that I encounter.

Lauren Doliva
Practice Leader
Global E-Business Practice
Heidrick & Struggles

Introduction

THE PROMISE OF YOUR HIGH-TECH CAREER

"The upside of money, handled correctly, is freedom."

—JEFF CUNNINGHAM, partner, Internet Finance Partners

The Bursting of the Dot-Com Bubble

The day I signed the contract to write this book, the Nasdaq Composite—the stock market index generally considered the best indication of the strength of the high-tech business sector—fell by 355.49 points. At the time, most observers thought (or hoped) it was a temporary glitch. But over the subsequent months, Nasdaq's slide continued, reflecting an unprecedented decline in technology stocks—the mirror image of the unprecedented rise they'd previously experienced.

As tech stocks declined, millions of people lost a lot of money (at least on paper). Hundreds of once high-flying Internet-based companies (the famous dot-coms) shut down, and many of the technology companies that had soared with them suffered serious declines in sales and profits. Business school graduates, who had been flocking to dot-com start-ups and other tech-related companies, began turning back to more traditional industries such as investment banking and consulting. And magazines that had featured the high-tech boom on their covers began running articles on the high-tech bust.

What does this shakeout mean for the future of careers in the technology sector? If you have a career in the high-tech world or are interested in pursuing one, are your dreams dead? Was the bright promise of technology merely an illusion after all?

The answer is an emphatic *no*. The high-tech employment landscape has clearly changed in ways that you need to understand, but as a career arena the high-tech universe is just as viable as ever, despite the hit it took. For the job seeker, *there's still gold in them thar' hills.*

During the technology gold rush of the late 1990s, growth in the sector was so wild that to get a dream high-tech job, an applicant basically needed only three things: a technical degree, the initiative to pick up the phone, and the ability to "fog a mirror." As Ross Brown, vice president of human resources for Analog Devices, puts it, "If you had a degree from a reputable school and you were a C student, you'd get three offers. If you were a B student, you'd get five offers. And if you were an A student from MIT, you'd get ten offers at $75K and up, with a 30 percent sign-on bonus."

The good fortune didn't end there. Those who hooked up with a business founder who had a bright idea and some venture capital funding had a great chance of making a fortune almost overnight. Hundreds of fledgling companies went public and found their stocks soaring in the atmosphere of high-tech mania. The company founders weren't the only ones who benefited. Employees with stock options often made millions as well. For example, at one time a stock run-up made millionaires out of 2,000 of Analog Device's 9,000 employees.

The influx of sudden wealth turned many people's heads. A CEO friend of mine who'd made $250 million in just a few years was driving down a California highway with her husband when they passed a billboard announcing that the top prize in the state lottery was up to $26 million. She told me, "We simultaneously looked at each other and shrugged, 'Is that all?' "

Of course, like any speculative bubble, the Internet stock craze could not last forever. When the dot-coms took a dive during 2000 and 2001, a lot of people ended up bankrupt. For many twenty-somethings, it was their first taste of career adversity. For others, who'd remained in traditional businesses and seethed jealously at the sight of their friends making millions (at least on paper), it felt like sweet revenge. The media discovered the German word *schadenfreude,* which means "pleasure derived from the misfortunes of others." It was used to describe the snickering heard from some quarters as Internet stock prices, such as those of InterWorld Corporation, tumbled—from a peak price of $93.50 per share to $2.94 in InterWorld's case.

A Longer Perspective on Dot-Com Mania

People who'd previously lived through two or three business cycles brought a realistic perspective to bear on the high-tech shakeout. After the giddy 1998–1999 period, the years 2000–2001 brought a temporary lull—in the stock market, in high-tech business growth, and in the job market. But this lull doesn't mean that the high-tech industries are dead. Far from it. The dot-com shakeout was a necessary correction, a return to reality in which hundreds of (often dumb) business ideas flopped—as they were bound to do sooner or later. The shakeout left the field free for the stronger, smarter high-tech companies, the ones truly contributing value to the U.S. and world economies through innovative business strategies and new technology-based products and services.

The collapse of many of the dot-com businesses built around retail Internet commerce—the so-called B2C (business-to-consumer) sector—was in reality a failure of marketing strategy. In fact, although these businesses were happy to hype themselves as "Internet companies" or "technology companies" in order to benefit from the mania for technology stocks, these weren't technology businesses at all. Their business model was simply a slight variation in the traditional "direct selling" business, which is as old as the Sears mail-order catalog of the late 1800s. Is the

Home Shopping Network an "electronics business" because it uses television as its sales medium? Of course not. And neither were the B2C dotcoms "technology businesses."

As we'll discuss in the next chapter, technology means computers, artificial intelligence, data management, fiber optics, data storage, microelectronics, telecommunications, satellite communications, anything wireless, and a variety of other extremely exciting areas in which new scientific breakthroughs are revolutionizing how we create, share, manage, store, and communicate information. These new tools are changing hundreds of traditional industries, from package delivery to home construction to travel and entertainment to auto manufacturing. But that doesn't make all those industries into high-tech industries. It's the companies that are developing and marketing the high-tech tools (and the services that go with them) that are true high-tech companies.

During the heady 1998–1999 period, many observers claimed that a "new economy" had been born, one that operated by new rules. And indeed there were a number of distinctive qualities to the business atmosphere of that period. There was an abundant supply of risk capital to fund the building of the "new economy" by entrepreneurs in start-up companies. There was a focus on revenues rather than on profits, with the assumption that profitability would follow in due course. And "old-economy" companies—sometimes referred to as "legacy" companies—responded to the perceived threat of the newcomers with huge increases in capital expenditures, often financed by borrowing.

All of these trends produced some predictable results. Many investors expected the new economy to be built at breakneck speed. Orders for equipment to build the new-economy companies (semiconductors, fiber optics, software, servers, routers, and so on) exploded, and chronic shortages appeared. Because of concern about these shortages, many companies ordered more than they needed.

All of this changed dramatically in the spring of 2000. The Nasdaq collapsed, led by the dot-coms. A new, much more somber mood of realism was ushered in. Risk capital for new ventures, including high-tech businesses, became almost nonexistent. Many dot-coms faced major funding problems, with many going bankrupt. The legacy companies that were scared out of their wits by the threat of the dot-coms responded with relief. Competing with the new-economy entrepreneurs had been expensive and unprofitable for them; with the bursting of the high-tech bubble, they could invest more slowly, more selectively, and more prudently. They cut capital expenditures aggressively.

What are the results of these changes in the business climate? The new economy will still get built, but over a much longer period, and largely by the old-economy companies that have the knowledge and capital to do it.

Companies that supply the equipment to build the new economy have experienced sharp slowdowns in orders. They can expect a much lower long-term growth rate. And as the increased information technology (IT) capacity now being built comes on line, pricing power will migrate from equipment suppliers to the companies doing the building.

The High-Tech Future

Technology still represents the future of our economy. Better yet, it can be *your* future. And the amazing developments of the last twenty years (starting with the first humble PCs of the early 1980s) are only the opening salvos in what promises to be a long-term explosion of new technologies. As Jerry Golley, CEO of AMI Visions, puts it, "Not only has the technology train not passed you by, it hasn't even started up the hill."

A few simple statistics indicate the continuing growth potential of the high-tech industries. Although the tech sector currently represents only 8 percent of the U.S. economy, it now provides 30 percent of the growth in the economy. During 2000, while the economy at large grew at a modest 4-percent rate, the high-tech sector experienced an increase of revenues of 68 percent. If you pick any period in the last decade, the odds are good that technology stocks led the market.

It's only natural that high-tech businesses are growing more rapidly than traditional industries. Their growth reflects a fundamental change in the nature of our economy. The *information content* of the products and services we buy has expanded enormously in recent decades. As software engineer Kevin Werner says, "Our society runs on information and data, and when technology helps us process that information faster or in a new way, there will always be lots of money to be made and jobs to be had."

Venture capitalists are seeing just as many new companies being launched today as in 1999. The difference is that traditional business measures such as healthy gross margins, a clear competitive advantage, a solid management team, and large potential markets can no longer be ignored, as they too often were during the days of dot-com mania. Investors are demanding a solid business rationale before funding high-tech ventures—as they should. As a result, the new technology businesses being launched today are likely to be better run, more stable, and more profitable than those of the generation of the late 1990s.

Where Are You Coming From?

Because you've picked up this book, I assume that you are either currently employed *outside* the technology sector and want to get into it, or you are employed *inside* the technology sector but want to accelerate your career growth.

Of course, there are some differences in the career strategies you need to employ, depending on which of these two situations applies to you. In this book, I'll discuss those differences in some detail. But there are certain strong similarities as well. In either case, in order to pursue your dream career in the high-tech field, you need to develop the skills to:

- Compete more effectively against other well-trained, highly motivated high-tech professionals.
- Make yourself into the leading candidate for the jobs you want.
- Communicate your value to employers in a professionally assertive and attractive manner.

- Move with confidence between companies, technical areas, and industries.
- Rise into the career levels where significant money can be made.

This book is designed to help you develop those skills. In researching it, I've called on dozens of high-tech players who've already found their version of the dream career: CEOs and executives from high-tech companies, executive recruiters, venture capitalists, and people known as centers of influence in the high-tech universe. I've quoted their comments and advice throughout the book. When you pay attention to these insiders' insights, the process of starting or building a high-tech career will be demystified and will become more accessible.

If you are currently employed in a non-technical industry, there's good news for you. You don't have to be a techie-nerd to get in on the high-tech excitement. SAT scores don't count for much either. You'll see that now is a good time for people from non-technical backgrounds to cross over to the technology world, and in this book you'll find specific tips and advice about how to do so. Yes, there is good money to be made in this exploding business sector. But more important, there is also an opportunity for unsurpassed excitement in building the future that you want.

The Money Factor

I've alluded to the fact that your earning potential is likely to be greater in the high-tech industries than in other, more traditional businesses. This is true, and for many people it's a very important factor in the decision to go high-tech. But earning millions is not the main motivation for everyone in the high-tech field, and does not need to be what motivates you. Having a good, fun, full, loving, balanced, and healthy life should be your goal. Lyle Hurst, director of HP World e-Inclusion for Hewlett Packard, puts it this way: "People I work with are looking for a variety of things: more solid footing in a fast-changing environment, more meaning in their work, more flexibility in when and where they work, more financial upside linked to their own contributions, and a feeling that they are not being left behind in the Internet revolution. Technology jobs offer more than high pay."

This is all true, and Lyle's words are an important reminder of some of the nonmonetary rewards that most people want from their work. You and I both know that money does not increase your worth as a human being. But money and a rich, meaningful way of life shouldn't be mutually exclusive. Used properly, money is a crucial ally in your pursuit of the good life, however you choose to define it.

Money gives you more time to do the things you want—to be with your family and friends, to travel, to give time to charitable and civic ventures, to enjoy creative activities or sports. Money can provide peace of mind by reducing your stress about your finances, your health, and your future security.

Money also helps you avoid having to work longer than you want, having to stay in a bad marriage or relationship for financial reasons, and having to delay or forgo altogether the pleasures of starting a family. As Jeff Cunningham, a partner at Internet Finance Partners, says: "The upside of money, handled correctly, is freedom."

In fact, the hard, cold truth is that sometimes the money you earn is the only permanent and reliable element in your life. Personal relationships don't always last; friends move away, parents die, marriages may end. Knowing that you're making money can be a powerful source of security in an insecure world. And perhaps best of all, it's something you can always share—with family, friends, or charitable causes.

This book is about pursuing a great life. For most people, that includes making decent money—a goal that you're more likely to achieve in the high-tech arena than in most other places in today's work world.

The Changing Image of the High-Tech World

Our image of the high-tech world is largely that it is a place where white males are thriving. When we think of technology, we picture magazine cover boys like Bill Gates (chairman of Microsoft), Michael Dell (founder and chairman of Dell Computer), Steve Jobs (founder, Apple Computer), John Chambers (CEO, Cisco Systems), and Steve Case (founder of AOL, chairman of AOL Time Warner).

Yet this white male image is an increasingly misleading one. The fact is that technology is the greatest equalizer of opportunity today. According to the research firm Venture One, women already hold 45 percent of the top-tier management positions in information technology, compared to only about 4 percent in traditional Fortune 500 companies. Carly Fiorina, the CEO of Hewlett Packard, has reached the very top of the high-tech hierarchy.

Barriers to success for women in high technology are gradually falling. Unfortunately, racial and ethnic minorities—specifically, Hispanics, African Americans, and Native Americans—haven't yet made their mark in high tech to the same degree. Studies suggest that several factors contribute to the lack of nonwhites in technology businesses, including:

- Lack of access to computers and other technological equipment at an early age
- Emphasis on violent games as the predominant function of computers
- The scarcity of teachers or counselors who are knowledgeable about the wide variety of career paths in technology
- The popular image of high-tech careers as solitary, antisocial, and intensely demanding ("You'll work 24/7!")
- Lack of encouragement for minority students to pursue degree programs in math and science
- The fact that technology isn't perceived as a "social conscience" field like law, medicine, or politics
- The relative paucity of high-tech role models for minority youth

If you're a member of a racial or ethnic minority, I urge you to pursue your high-tech dreams. High-tech companies in which minority-group employees play leading roles will be more likely to develop products and services that respond to the needs and interests of minority markets. For example, the so-called "digital divide" that prevents many inner-city schools from having access to the full power of modern information technology is more likely to be overcome when minority group members have key decision-making roles at the leading software, hardware, and infrastructure companies. And, of course, minority groups will benefit from being able to share in the business and economic growth that the high-tech world will experience in the decades to come.

Large and powerful as it may appear, today's high-tech industry is still in its infancy. Many of the great companies that will dominate the technology business twenty years from now are unknown today. Now is the time for women and minority group members to get in on the ground floor, so that they can occupy their share of the boardroom chairs in the not-too-distant future.

What You'll Learn from This Book

Here's a brief overview of the contents of this book and what you can expect to learn from each chapter.

Chapter 1 examines the *broad range of technology industries,* offering statistics and information about which businesses and careers are included in the high-tech universe and their prospects for growth.

Chapter 2 offers guidance about *how to decide whether to stay in your current job or go elsewhere.* It covers such trends as internal job shifts and rehiring of departed employees, and offers tips on dealing constructively with job loss.

Chapter 3 shows you *how to set your compass.* You'll learn how to capture on paper the who, what, when, where, and how of your life so as to raise your own self-awareness, identify the skills and talents you have to offer, and clarify your goals for life and work.

Chapter 4 is about *polishing your product*—improving the skills you need to be a valuable commodity in today's high-tech job market. This chapter explores the importance of continual learning (and shows how to get it), and describes the combination of old-economy and new-economy skills and attitudes that the most successful people in high tech are using.

Chapter 5 will help you *make yourself known.* Leaps in job advancement evolve around personal sources that you develop. This chapter is about how to make it happen through networking techniques that anyone can use, and that everyone needs to know.

Chapter 6 will show you how to *launch your job search.* The chapter explains many of the best ways of researching and reaching out to companies where you may want to work. It also describes the advantages and disadvantages of the online job search.

Chapter 7 covers *working with headhunters*—the business consultants who handle an increasing percentage of the most lucrative and important high-tech jobs. You'll learn what you can realistically expect from a headhunter and how to get the greatest benefits from working with one. You'll also learn about a little-understood additional source of high-tech jobs, the venture capital firms that finance many new-economy companies.

Chapter 8 will help you *prepare the tools you'll need* for your job search—your resume, cover letters, bio, and other supporting documents. You'll also find proven tips for preparing for high-tech job interviews, including how to handle the most commonly used interview questions.

Chapter 9 is about *evaluating job opportunities.* Before signing on with any company, it's important to find out all you can about the organization, its financial status, its management style, its culture, and its future prospects. In this chapter, we'll show you how to uncover the truths behind the rosy pictures recruiters like to paint.

Chapter 10 will guide you in *negotiating your deal.* You'll learn how to handle discussions concerning salary, benefits, and perks in a style that's both aggressive and professional. You'll also learn about how to get the best possible raise or promotion from your *current* company, whether or not you've been looking for new opportunities elsewhere.

Chapter 11 will show you *how to go for the gold* by identifying and pursuing opportunities for achieving real wealth in high-tech business. It examines the potential in consulting and entrepreneurship as well as the sometimes complicated ins and outs of employee stock options.

Chapter 12 provides tips about how you can *hit the ground running*—how you can get off to a great start during your first thirty days at a new job, days that largely determine how successful you will be throughout your tenure with the company.

The Epilogue offers enlightening insights from high-tech professionals who've "been there, done that," and now share their hard-earned wisdom, telling you what they would have done differently in their careers if they'd been smarter sooner.

Does this all sound enticing? I hope so, because the opportunities we'll explore together are truly exciting ones.

From Biotech to Hollywood

1

THE EMERGING HIGH-TECH UNIVERSE

"Tattoos, Nose Rings, and Pin Stripes, Welcome . . . Come as you are."

—Classified job ad during the high-tech boom, *The Wall Street Journal*

Defining High Technology

It's hard to define in just a few words exactly what the high-tech sector encompasses. This is true partly because the technologies themselves are constantly evolving and diversifying, and partly because the uses to which they are applied are constantly multiplying. There is virtually no major industry today still untouched by high technology.

High tech is also hard to define because the technology itself is often dauntingly complex. Insiders at Microsoft say that the most difficult challenge in marketing any new technology is figuring out how to explain it to others. When Microsoft.NET was introduced, a host of executives at the company were interviewed about it. Nearly all had difficulty explaining how the new product worked. "Give them an hour and you began to get an idea, but even then I felt like one of the blind men feeling the elephant," wrote a *Fortune* magazine reporter. "No wonder the public announcement in June took four hours," Steve Ballmer, CEO of Microsoft, had to admit, "the last big memo we put out about this is already out of date."

One simple, down-to-earth explanation of high tech is this one: "It's anything that makes stuff smaller, faster, and more powerful, making it possible to do more things more efficiently and more easily while using fewer people." But in more concrete terms, "technology sector" refers primarily to computer-based information systems and technologies that support and employ them, including computer hardware and software, associated peripheral devices, design and production of microchips, Internet commerce, telecommunications (including voice, data, and video), and computer applications in a host of industries, from manufacturing to agriculture to health care to education.

To further illustrate the depth and complexity of today's high-tech sector, just peruse the following (partial!) list of subsidiary fields in some of the major technology areas:

COMPUTERS

Artificial intelligence

CAD/CAM

Computer leasing

Data management

Computer hardware design and manufacturing

Networking

Personal digital assistants

Network security and encryption

Servers

Software applications

Internet appliances

IT consulting services

ELECTRONICS AND SEMICONDUCTORS

Electronics components

Electronic display devices

Microelectronics

Optoelectronics

Microwave electronics

Lasers

TELECOMMUNICATIONS

Broadband

Cable industry

IP telephony

Local and long-distance carriers

Satellite communication

Telephone networking

Wire and cable

Wireless

Cable modems

Digital subscriber lines (DSL)

INTERNET

Application Service Providers (ASPs)

Internet Service Providers (ISPs)

E-commerce

E-mail

Infrastructure and site management

Web site development and design

Web hosting

Electronic Data Interchange (EDI)

Mobile commerce

Customer relationship management software

E-fulfillment

Web store solutions

BIOTECHNOLOGY

Bioinformatics

Cloning

Gene therapy

Genomics

Protemics

Stem cell research

Lists such as these become outdated almost as soon as they are compiled. A single recent issue of MIT's magazine *Technology Review* wrote about such emerging technologies as speech, image, facial, and fingerprint recognition; microfluidics (using liquid particles a thousand times smaller than a drop of liquid); microphotonics (using tiny crystals to reflect light, which moves bits of information more quickly through the Internet); aspect-oriented programming (which enables new kinds of shortcuts in writing computer code); microchips made from plastic or organic material (rather than the traditional silicon); and new applications of robotics. Each of these technologies is potentially the source of dozens of new companies. In fact, robotics alone may well represent a host of new industries by the time the first decade of the twenty-first century is over. "Robots [today] are where PCs were in 1980," says *Technology Review* editor David Talbot. "[They're] poised to break into the marketplace as common corporate tools and ubiquitous consumer products performing life's tedious chores."

Every Business Is a High-Tech Business

New information technology (IT) is having an impact on virtually every business sector: financial services, retailing, insurance, food and beverage production and marketing, advertising, aerospace, energy and utilities, agriculture, manufacturing, real estate, construction, travel and leisure, chemicals, health care, pharmaceuticals, entertainment, media, automobiles, publishing, defense. . . . The list is endless. And within these industries, almost every business function is becoming increasingly digitized, from sales and marketing to finance, accounting, and administration.

Think "high-tech business" and your mind quickly turns to companies such as Microsoft, Intel, IBM, Apple, Compaq, AOL, Yahoo!, eBay, Cisco, and Oracle. But high tech also includes businesses such as these:

- Burlington Northern and Santa Fe Railway Company, which uses Internet-based auction technology to sell locomotives
- Telephone companies, where line workers use laptop computers to install and repair residential phone lines instead of searching the neighborhood for a phone box on a telephone pole
- American Honda Motor Company, whose Web-based supply chain allows 1,250 dealers to order specific models online, drastically reducing delivery times
- Bricklaying, where bricklayers doing repair work use e-mail and digital cameras to send architects photographs of problems
- Safeway grocery chain, which uses a pricing code system that electronically changes prices without anyone touching the products on the shelves
- Airlines, where airplane engine repair manuals once contained in six-inch-thick binders now fit on a single easily updated CD-ROM

Amazon.com, the online retailer of books, music, and a growing list of other products, is one of the world's most advanced e-companies. But another is Wal-Mart, which has used the new IT to create the world's most progressive and powerful backend system for managing inventory, logistics, working capital, and customer data.

Technology is becoming ubiquitous. As Thomas Weber, e-world columnist for *The Wall Street Journal*, writes, "Look around at the conveniences of everyday life. Check out your car, your cell phone, maybe your wristwatch. Can you tell what's a computer and what's not? The line is increasingly blurry. Computer chips and software have made their way into everything from treadmills to toasters."

Thousands of companies are already using new IT to improve both their external operations (by enhancing customer service via the Internet, for example) and their internal efficiencies (by using a company intranet to improve manufacturing quality, for example). And those that are not already deep in the throes of this kind of upgrade are generally involved in exploring where to begin, usually with advice and guidance from a host of consultants and advisers. One such adviser—Charles Wallace, a managing partner at the human resources consulting firm of Heidrick & Struggles—says, "The challenge is the confusion about the implication of technology for their business—how to use it better, how to compete more effectively. It takes constant preaching, educating, and consulting to get them to upgrade talent, for example. If they don't enlighten themselves, they will lose market share." More and more companies are getting the message—which means the creation of still more jobs in the high-tech sector.

The Internet Revolution

A single sector of technology, the Internet, was the biggest business story of the late 1990s and early 2000s. It may be the biggest story throughout the rest of this decade as well. As high-speed Internet connections capable of transmitting voice, data, and images anywhere proliferate, the applications of the Web will continue to multiply, revolutionizing business, communications, entertainment, education, politics, and almost every other field of human enterprise.

What makes the Internet so significant is the incredible breadth of applications it supports. Chris Vargas, the CEO of F-Secure, explains it like this: "The Internet connects on a business, emotional, and financial level. Everything is connected to the Internet: Computers, disk drives, software, fiber, routers, and switches are all connected. Even in low-technology businesses, like a car, with GPS, you now have wireless technology." GPS (global positioning system) refers to the use of satellites to track the location of individual vehicles anywhere on the surface of the earth. Combining the information provided by GPS with the communications power of the Internet, auto companies can now offer instant driving directions. (Similar technology is already used by overnight delivery companies as well as firms such as Exxon, which operates a fleet of oil tankers that often must be rerouted in mid-shipment.)

GPS is yet another remarkable example of how new IT, facilitated by the Internet, is remaking how everyday activities are conducted. No wonder Chris Vargas concludes, "The Internet transforms the way we do everything."

The Tech Sector Job Boom

Clearly, the high-tech revolution is having an impact on business way beyond the computer hardware and software companies. Firms in every industry are busily working to figure out how to use the new technologies to improve their own operations, to serve customers better, and to gain a competitive edge. Many companies are launching new technology-oriented divisions, expanding traditional IT departments, creating spin-off companies devoted to high-tech opportunities, and adding tech-savvy employees to existing operations. And every day new high-tech companies are being founded (though at a less frenzied pace than during the era of the dot-com boom). As a result, there's enormous demand for people trained in high technology, or even for those simply comfortable in working with high-tech products and services.

Consider the following current statistics, compiled from various industry and government sources:

- Sixty percent of companies report that they need more employees to meet their current operational needs in information technology.
- 350,000 tech jobs are currently unfilled in the United States.
- The fastest-growing job opportunities in the United States are for computer engineers, computer-support specialists, and systems analysts.
- Eighty-three percent of companies plan to increase the numbers of IT workers on their payrolls in the coming year.
- 800,000 IT jobs will go begging for lack of workers in the next two years.
- By 2006, the United States will need 1.3 million new high-tech workers.
- IT job openings for new college graduates rose by more than 40 percent in 2000.
- IT jobs pay on average 20 percent more than comparable nonhigh tech positions.

It's obvious that the boom in the high-tech sector has created a classic seller's market for technical professionals. If you have high-tech skills—or are willing and ready to develop them—you will be much in demand in the job market, today and for years to come.

What are some of the specific positions you can find? They're almost too numerous to mention, but let's consider some examples. A technology company needs technically trained individuals to work as:

- **Engineers** in design and development, software, hardware, compliance, manufacturing, process, quality, integration, packaging, performance testing, diagnostics, product design, software testing, applications, systems architecture, server development, configuration management, and more
- **Technicians** trained in troubleshooting, fixing, and supporting all of the above functions
- **Managers** in all of the engineering disciplines

- **Information technology professionals,** such as implementation project managers, system implementation engineers, technical writers, software engineers, web commerce developers, network security specialists, IT managers, software analysts, software programmers, programmer analysts
- **Managers** for all of these professionals

The same company also needs nontechnically trained individuals for the following jobs:

- **Managers** in business development, contracts administration, electrical, materials, information systems, operations, production, purchasing, quality assurance, supply base, and material operations
- **Financial professions,** including auditors, controllers, cost analysts, stock administrators, tax specialists, accountants, purchasing managers
- **Administrative and human resources professionals,** including HR representatives, recruiters, benefits specialists, training specialists, course developers, general counsels, contract administrators, international attorneys, licensing experts, real estate managers
- **Operations professionals,** including operations directors, managers, technicians, shift supervisors, buyers, order administrators, export compliance managers, traffic specialists
- **Marketing and sales professionals,** including marketing directors, account executives, market researchers, pricing and policy specialists, product managers, data analysts, sector strategists, marketing communications experts, trade show specialists, promotions specialists, public relations professionals, merchandising managers, sales coordinators
- **Executive team members,** including vice presidents, the president, the CEO, and the chairman
- **Executive assistants** for all of the above

Types of Companies

Considering the job options available, you can choose to work in the high-tech field for any of several types of companies, including:

- Large blue-chip technology companies (such as Sun, Oracle, and Microsoft)
- Small- to medium-sized single-technology companies (such as Arriba, FirstData, and Selectron)
- Start-up technology companies (such as BrightPlanet, siteROCK, and bITssol)
- Large nontech companies with IT requirements
- Smaller nontech companies with IT requirements
- "Outsource" firms that provide consulting services to all of the above companies

As you can see, you have a lot of choices to make. Throughout this book, the pros and cons of various kinds of companies and jobs will be examined. For now, let's take a look at portraits of two extremes: a small start-up technology company and a large blue-chip legacy company. Most companies will fall somewhere between these two extremes. As you read the descriptions that follow, try to imagine what it would be like to work for either of these imaginary companies. Which place feels more like home to you? Where would you be more comfortable? The answers to these questions can help you start thinking about your options and guide you toward certain types of companies and away from others, depending on what appeals to you most.

THE SMALL START-UP TECH FIRM

This type of company is likely to have:

- A sense that no one is really in charge
- A sense of chaotic growth (say, from a staff of 14 to a staff of 400 within 12 months)

- Lots of wasted money (usually the investors' money)
- No time to train you
- Little stability
- Limited support resources
- Fourteen-hour days, seven days a week
- Zero vacation or free time
- No clear payroll or human resources policies
- No clear practices
- No one to complain to
- No experienced managers as mentors
- More openness between management and rank and file
- Increased speed of project execution
- Shorter time for advancement
- Increased risk of failure

THE BLUE-CHIP LEGACY COMPANY

This type of company is likely to have:

- Opportunity at almost every level
- Resources
- Profitability (more likely, but not guaranteed)
- A proven reputation
- Security
- Stability
- A large peer group of colleagues
- Experienced managers as mentors
- Assistants
- Vacations

- Longer wait for advancement
- Greater chance of ultimate success

We'll take a look at the differences between various kinds of companies in the high-tech arena in much more detail later in this book. But it's not too soon for you to begin thinking about the type of working atmosphere you think you'd most enjoy and in which you're likely to be most productive and successful.

Of course, you may already be a veteran of the job wars—an experienced working person with a history at one or several companies. If so, you already have a sense of the kind of business environment that suits you, based on both good and bad experiences from your career. And one of the immediate career challenges you may be facing is whether or not you should leave your present position or stay where you are. The next chapter will help you answer that question.

2

SHOULD I STAY OR SHOULD I GO?

"Climbing the corporate ladder doesn't typically occur by staying at a single company anymore. It's about moving around to gain a broad base of experience across industries."

—KATHERINE CIZYNSKI, partner, Wiser Partners

The New Employment Ideal

There was a time—perhaps in your parents' generation—when the typical American worker dreamed about lifetime employment at a single great company. In the eyes of millions, the ideal career consisted of joining a large organization right out of school, starting at the bottom, and working diligently and loyally for the same firm, rising gradually through the ranks, then retiring with honors from the post of vice president or regional manager and enjoying a well-deserved company pension at age sixty-five.

The downside to this image of reciprocal loyalty between employee and employer was the stigma attached to unemployment. For anyone out of a job, whether because of a large-scale company layoff, an individual firing, or even by choice, it was widely assumed that there must be something wrong with that person. And someone changing jobs with any frequency—even three or four times in a decade—would probably be branded a "job-hopper" and regarded as a bad employment risk.

Today, the old ideal of lifetime employment is long dead. Many factors contributed to its demise, but the overriding reason is simply *change.* The business world of the twenty-first century is far more competitive, dynamic, unpredictable, and fast-paced than that of the 1950s, 1960s, or 1970s. As a result, companies insist on the flexibility they need to respond quickly to changes in the environment, including new technologies, new competitive threats, and new economic trends that can destroy companies overnight and create new ones just as fast. And this flexibility includes the freedom to hire and fire employees, move locations, shut and open divisions, merge and spin off companies, and retool company processes at a moment's notice. When companies take these steps, the impact on workers is enormous.

In response, workers have adapted a free-agent mentality. The typical employee no longer assumes that his or her job is for life. Instead, the job is viewed as a temporary arrangement, one that will last as long as it's in the best interest of both employer and employee. Both are free to make a change whenever they consider it advantageous.

Just as the old idea that a big company "owes" its workers long-time employment is gone, and layoffs, firings, and downsizings are no longer viewed as corporate crimes, but as smart and necessary business practices, the stigma of unemployment or job-hopping has largely disappeared.

The new realities of job tenure can be illustrated statistically. For example, the Bureau of Labor Statistics reports that the median number of years people stay with an employer is now just 3.6. Thus, if you have a 40-year working life, you can probably expect to work for eight to ten companies during that time.

This can also be illustrated anecdotally. In your parents' generation, the stories swapped by successful business people often involved moving up through the corporate ranks of a single organization. Today, the stories more often turn on well-timed, thoughtfully planned hops from one company to another.

Tom Butta enjoyed a highly successful stint at Young & Rubicam, the great advertising agency. He used this experience as a springboard to start his own agency, which grew to more than $100 million in annual billings. Tom sold the business, then became the marketing chief of technology at Red Hat, a start-up company that markets Linux-based software. Tom's moves between companies and even between industries have been enormously beneficial to his career. He says, "The risks [of making a switch] are not as great as one might think."

Joan Gustafson, a vice president at 3M, has observed a difference in the career fortunes of those who job-hop and those who stay put. She says, "I found that most people who have moved from company to company were promoted each time they made a move. Those who stayed with the same company for all of their careers did not rise to as high a level as those who changed companies." Today, it's not uncommon to hear stories about someone making $60,000 in one company who accepts an offer from a competitor at $80,000, only to be hired back later by the original company for $100,000.

It's enough to make you conclude that job-hopping is the new way of life. It seems to be particularly prevalent in Silicon Valley and in other strongholds of the new economy, where the rise or fall of individual companies and of various ever-shifting technologies helps produce industry upheavals on a regular basis. It's certainly true that people who would once have been frowned upon as job-hoppers are no longer criticized for their moves. In fact, someone who stays in a single job for twenty years or more is more likely to be criticized or be considered "too conservative" or "a stick-in-the-mud." Job movement is certainly the new norm.

This isn't to say that changing jobs as often as you change your socks is considered smart career management. One company per career is no longer expected or even desirable. But a new company every year isn't usually desirable either. Finding the right balance has become a challenge for every working person—something only you can decide for yourself.

The Retention Dilemma

Of course, the new career flexibility is creating challenges for companies as well. As more and more workers become accustomed to the idea of jumping ship whenever it benefits them, more and more companies are struggling to retain their best employees.

When employees quit, it can be very costly for the companies that lose them. The knowledge, experience, and expertise they take away is often irreplaceable. And in addition to the cost of recruiting, hiring, and training new workers, companies have to deal with the poor employee morale often found in the wake of one or more departures. Often, employees who haven't moved start to review their own worth, become disgruntled about having to pick up some of the workload, and start to wonder whether they too ought to leave. They start asking themselves troubling questions, such as: Am I advancing fast enough compared to others? Will staying in this job hinder my future advancement, particularly if all the "good" people around me are leaving? Will having too long a tenure with one company make my resume look weak? Such employee concerns can cause a handful of defections to turn into a mass exodus.

This isn't just a theoretical problem. For hundreds of companies, especially in the high-tech arena, retention has turned into a critical success issue. Companies are trying all kinds of ploys to keep their best workers in the fold. One company in the San Francisco Bay area was offering new MBA hires $1 million if they stayed with the company for two years.

Companies have long recognized the importance of smart employee recruiting. Now they are discovering that it's just as important to spend their time on retention. This may involve financial factors, of course. AON Consulting studied 1,140 high-tech workers and found that 56 percent would stay at their current job longer if offered a similar job with slightly higher pay. But issues related to job satisfaction are at least as crucial as financial incentives. "You rehire your employees every single day," says Gloree Parker-Roden, senior vice president of Enterprise Services at Pearson Technology Centre. "We had no problem with retention until a time period where we went into a maintenance program without a major initiative going. Things felt slow to people. We found that the same people who complained about long hours and overwork really didn't want to underwork either." In other words, high retention of valuable employees requires keeping them challenged and interested as well as rewarding them financially.

In a research project, Watson Wyatt Worldwide found that the number-one driver of employee commitment is trust in senior management. If a manager fails to provide the necessary leadership, then people leave. Gloree Parker-Roden speaks for many when she says, "The important thing to me is being able to work with managers I respect and trust. If that was broken, then I would go."

The Watson Wyatt survey identified other factors that caused people to leave an employer. Here they are, starting with the most commonly cited:

- Higher salaries offered by other organizations
- Dissatisfaction with potential career growth
- Feeling unappreciated
- Rising acceptability of job-hopping
- Difficulty balancing work/life issues
- Burnout
- Benefits offered by other organizations
- Perceived lack of job security
- Conflicts with supervisor or co-workers
- Viability of the organization
- Conflicts with the organization's mission or values

Salary is clearly an important factor, but it's far from the only one. Any manager who has a high turnover rate shouldn't blame the company pay scale alone; instead, the manager should examine his or her own practices and the work environment.

What about effective retention tools? Wyatt Watson found that the best way to keep good employees was by offering attractive health care benefits and a competitive salary program. Other powerful tools included competitive vacation/holiday benefits, regular salary reviews and increases, an attractive retirement plan, paid personal time off, flexible work schedules, opportunities for training and development, a genuine open-door policy on the part of management, a strong orientation program for new employees, help with child care, tuition reimbursement, retention bonuses, and stock options.

You may have little say in designing your company's retention program, but the preceding list is a good starting point in considering the kinds of perks you want to ask about when evaluating future job offers.

Greener Grass—Leaving for the Right Reasons

So there can be many benefits to switching jobs. Moving to another company is likely to bring you a bigger and quicker raise and perhaps an opportunity for promotion. And many people find it stimulating and refreshing to periodically take on new challenges, new surroundings, and new colleagues. There's an excitement about changing jobs that can be a bit intoxicating. The first few weeks at the new company often feel like the early days of a new romance. You're "the new kid on the block" and everyone is curious about you and interested in you. You and your new co-workers haven't yet had the chance to learn about one another's shortcomings. Everything you say and do is observed hopefully and optimistically.

By contrast, "sticking it out" in a job you've held for some months or years can feel stifling, boring, and oppressive. The excitement of the early weeks has long since vanished, and in its place is routine work that often seems like drudgery. You've had a chance to discover your company's and colleagues' flaws, and they've learned about all of yours. As a result, it's easy to start daydreaming about the next job.

We all have days when we want to "chuck it all" and escape to a new job—*any* job. But that feeling alone usually isn't a good reason to do it. Remember, job changes have their downside, too. There's always a cost and risk involved in shifting companies, industries, and career tracks; it's never possible to be absolutely certain that your new position will really be better than the familiar one you're leaving. And there's a definite psychological, emotional, and physical cost attached to the personal upheaval of changing jobs. Don't get me wrong—I'm all for change with a *purpose.* But you need to really understand your motivations in order to make an intelligent and successful job switch.

The best way to decide whether to stay or go is to ask yourself some serious questions that can help you focus on whether or not leaving your current job makes sense for you. The following quiz is designed to help you do just that.

THE STAY-OR-GO QUIZ

Directions: Indicate how strongly you agree or disagree with each of the following statements. Choose your answers from the following scale:

5 = Strongly agree

4 = Agree

3 = Partially agree, partially disagree

2 = Disagree

1 = Strongly disagree

1. My job offers me real opportunities to expand my career horizons and job skills.
2. In my job, I spend at least half of my time doing work I'd like to do more of in the future.
3. My job offers significant upside potential for advancement.
4. My job offers me the opportunity to meet and work with interesting people from whom I can learn a lot.
5. My job gives me the opportunity to manage my own work and take responsibility for the results.

6. The people I report to know and value me and my work.
7. The people I report to are known and valued by those to whom they report.
8. Most days, my job is fairly interesting.
9. My job lets me live and work in a geographic location that I am happy with.
10. I am paid the kind of salary that I feel is fair and that satisfactorily covers my personal needs.
11. There are significant opportunities for advancement in other departments or divisions of my company.
12. My employer offers benefits (health care, vacations, retirement plan, etc.) that meet my needs.
13. My working conditions are reasonably comfortable for me.
14. I like the products and services my company produces.
15. I am proud to be associated with the policies and ethical standards that my employer represents.

How to score: Add the ratings on all fifteen questions. If the total is 50 or more, your job situation is basically a satisfactory one. If the total is between 35 and 49, there are significant problem areas you need to address. If the total is 34 or less, your job is basically unsatisfactory, and you should seriously consider the possibility of moving on.

The Stay-or-Go quiz should help you focus on the most basic reasons for considering a job change. But whether you choose to stay or go, do it for the right reasons. Don't do it just because "everyone else is doing it," because you've been in one place for three or four years, because it will look good on your resume, because you can't get along with certain people you work with, or just because your husband, wife, or best friend is pushing you to move. Don't do it for any reason that does not fit your long-term career goals.

The Internal Search: An Alternative to Leaving

Sometimes, changing companies isn't the best alternative, even for people who truly need a job change. If you think you fit with your current company but aren't being adequately challenged or rewarded in your present job, it may make sense to consider moving within the company. Don't assume this is impossible. If you seriously investigate the opportunities within your current company, you may be pleasantly surprised by what you discover. Dale Telford, former IT director at StarChoice, and now the founder of bITssol, puts it well. He suggests, "If there comes a time where you feel you can't go any further at your current company, talk with your superiors and let them know what you feel you're capable of doing. Ask them for suggestions or if they know of a position open in another organization. That way, if you do find something outside of your current company, you will not be surprising anyone. You may also find that even though you did not think there was something else you could grow into, the company you currently work for may know your real value and create the position you want."

Because keeping good employees has become a high priority at most smart companies today, your employer may be willing to facilitate your job change within the organization. Avoiding losing you to an outside competitor will help the company avoid the costs of recruiting, hiring, and training a replacement. There are benefits for you as well. You and your family will undergo less of a disruption, and the knowledge you've developed about your company, the people who work there, its systems and processes, and its customers and competitors will all continue to be useful to you.

A lateral or upward shift within the same company may be the ideal move for you. Even if the new department or division doesn't turn out to be a perfect long-term career match, the new assignment could turn into a useful "bridge" job that keeps you sane and teaches you some new skills as you consider other options.

If you're interested in pursuing an internal change, start by talking to your boss. Think carefully about how to approach the subject. Use tact and diplomacy to explain why you're interested in a change without expressing bitterness, anger, or boredom with your current job. The last thing you want to do is to provoke defensiveness or hostility—after all, you want your boss's help in making the shift. Talk in terms of your aspirations for the future rather than emphasizing your disappointment with the past and present. Don't say, "I want a new job because the work here is depressing, dull, and pointless." Instead, you can say, "I think I'm ready to tackle some new challenges at a little higher level of responsibility that will benefit the company."

It's unlikely that your boss will be in a position to directly link you with job opportunities in other departments; your company's human resource or personnel department will have to play that role. However, company policy usually requires your boss's approval for an internal job search. Further, it would scarcely be comfortable to proceed against your boss's wishes, which is why it's a good idea to start with him or her. And who knows? Your boss *may* surprise you by responding, "I had no idea you felt that way. Would you like to be considered for the new job that's just about to open up in our department?"

When the Choice Is Out of Your Hands

Unfortunately, as much as we'd like to think that our careers are entirely in our hands, there are many factors over which we have no control. Keeping your current job may be one of them.

Of course, there are things you can do to increase your job security. Working hard, continually learning, and taking responsibility are all critically important. So are people skills, such as getting along with colleagues above, below, and alongside you; being a positive, encouraging presence; and dealing with others honestly and tactfully. Also important is professional self-promotion—that is, making sure that your talents and accomplishments are recognized appropriately, without being boastful or pushy.

But even when you do everything right, your job may still be in jeopardy. Cynics say, "Your job is only as secure as the emotions of your immediate supervisor," and while this is an exaggeration, it's true that bosses sometimes let employees go for reasons that are more related to their own managerial or psychological shortcomings than to the failures of the worker. Further, you may be doing superb work in a company or an industry that is floundering, or you may be caught in a numbers game where in order to meet Wall Street's profit expectations, the company decides it has to cut costs by 10 percent across the board, which means that somebody has to walk the plank—it hardly matters who.

So you can do *everything* right, and things can still go *splat*. As John Elway, two-time Super Bowl champion quarterback, says, "Not only do you have to be good, but you have to be *lucky*."

Of course, it helps if you can anticipate when your job is at risk. This can enable you to prepare for the worst, not only in terms of your job search (updating your resume, expanding your networking), but emotionally and financially as well. You can get psychologically accustomed to the idea of switching companies, you can speak with your spouse, family, or significant other about the possibilities, and you can increase (or start) your rainy-day saving and investment plan, just in case you need to survive for several months without a regular paycheck.

SIGNS THAT YOUR JOB MAY BE AT RISK

Look for some of the following warning signs to determine whether your current job is at risk.

- You sense disregard for your authority by those above, below, and around you.
- The number of people who report to you is significantly reduced.
- You increasingly feel that your energy, enthusiasm, and smarts are getting you nowhere.
- You experience notable indignities, such as being ignored in meetings, being left out of the loop on key decisions, or being omitted from the circulation lists for important memos or e-mails.

- You have frequent run-ins with peers.
- You are repeatedly passed over for the most interesting, important, or prestigious assignments.

WARNING SIGNS OF TROUBLE FOR YOUR COMPANY

Sometimes it's the company itself that faces an uncertain future. This, of course, can have a significant impact on employees. Consider the following signs of a troubled company:

- Partnerships and co-marketing deals with other companies are canceled.
- Two or more leading company executives resign or are fired.
- One or more well-known figures resign from the board of directors.
- Anticipated rounds of capital funding are reduced or canceled.
- The stock price takes repeated hits.
- There are spending and investment cutbacks.

One survivor of the corporate wars summarizes the downward slope of bad news like this: "First the bigwigs say, 'We're in transition.' Then they say, 'We're downsizing.' Finally they say, 'We're closing.'"

Moving On

Whether it's "your fault" or not, being let go is a traumatic experience. It invariably produces emotions ranging from disbelief and rage to guilt, shame, and depression. It generally creates financial stress, which can lead to serious anxiety and conflicts within families. And the sense of uncertainty and loss of control that goes with being fired can make it more difficult to manage a job search in an intelligent, energetic, creative fashion.

There are entire books that focus on the trauma of job loss, along with career counselors and psychologists who specialize in helping layoff victims cope and regroup. Clearly, all the issues associated with job loss can't be addressed in a few pages. Nevertheless, here are a few suggestions.

First, remember that in today's job market, the stigma of being let go is much, much less significant than ever before. The continual waves of mergers, spin-offs, company launches and closings, downsizings, expansions, and re-engineerings that have marked world industry over the past two decades mean that *almost everyone* has been laid off at one time or another. And those who haven't been fired are well aware that they've escaped the ax only by good fortune. As a result, no one really looks down on people with one or two layoffs in their past. Instead, layoffs are viewed as par for the course. Try mentioning being fired the next time you're at lunch with half a dozen other working folks. Rather than glances of disapproval, you're more likely to see nods of understanding and hear comments like, "I've been there." Of course, this doesn't apply if you've been fired repeatedly, or for a cause like lying, stealing, or punching your boss. But most people who are one- or two-time losers in the job wars have little difficulty moving on to their next opportunity.

Second, you can minimize the psychological and career damage of being laid off by handling the process intelligently. Here are some of the steps to take:

- Negotiate a fair severance package. Rather than simply accepting the company offer, request a couple of days to consider it. Then talk to peers and former bosses and colleagues. You may find you can get the company to increase your final pay package, extend your health and life insurance benefits, or provide you with services such as career counseling.
- Try to analyze objectively what you did right and wrong. Examine your history with the company. Could you have handled your job better so as to extend your tenure? Were there warning signs you ignored? Make a list of work and people skills you intend to improve in your future jobs.

- Start your next job search promptly. Don't spend weeks watching TV or feeling sorry for yourself, or spend your savings on a vacation or some other consolation prize, even if you can afford it. The loss of psychological momentum you'll suffer can be harmful.
- Don't be embarrassed about losing your job. Develop a simple, neutral, accurate, one-sentence explanation for why you lost your job. For example, you can say "The company reorganized, and mine was one of several positions that were eliminated." Using this or a similar sentence to spread the word among your friends and acquaintances will maximize your chances of hearing about a worthwhile opportunity.

The great automobile entrepreneur Henry Ford said, "Failure is the opportunity to begin again more intelligently." His words apply to getting fired. Don't waste time and energy bemoaning what you've lost; focus instead on the new horizons in your future.

You *Can* Go Home Again

Another effect of the changing business landscape has been a change in the attitude toward departed employees. In the past, many companies treated those who'd left (whether voluntarily or involuntarily) as pariahs, traitors, or losers. Today, they're generally considered a pool of potentially desirable rehires. After all, if you leave Company A for a job at Company B, you will still retain useful knowledge of how Company A operates. In addition, you will also develop valuable new insights into the strengths and weaknesses of a competing firm. Thus Company A will probably be happy to consider hiring you back. Rather than a "traitor," you're more likely to be called "a returning alumnus."

Hiring alumni is good for both parties. A returning employee requires less orientation and retraining than someone completely new, leading to a shorter transition time and lower hiring costs for the company. The employee gets a proven, familiar infrastructure to return to and often comes back equipped with a clearer view of what can and should be in his or her career mission.

Many companies recognize the value of rehiring former employees. In some divisions of GE, for example, departing employees are officially put on an "unpaid leave of absence," which makes it easy for them to return within a few months in case they change their minds. At Bain & Company, a major consulting firm, defectors continue to be invited to company social gatherings. Some companies use alumni for new-hire referrals and feedback on employee retention. In return, they may offer free career advice, the latest industry/company news, and even free office space.

Accenture recently sent a letter to 3,000 of the firm's recently departed alumni telling them the door was open if they wanted to come back. The firm also stays in touch with alumni as potential clients, because many former Accenture consultants become senior managers at other companies. Even some dot-com companies who've lost people have sent letters to alumni offering them extra equity if they return before a certain date.

A marketing manager—Chase—left that position with one company for a vice president position elsewhere. His original employer, having failed to fill his slot successfully, called Chase six months later in an effort to woo him back. "Okay," Chase responded, "but I'm a vice president now." Although his recruiter hadn't planned on it, Chase got that title.

Your Best Bet: Stay Employed

If possible, stay employed while job hunting. Although it's more important to be *employable* than to be employed, there is still a positive bias toward someone currently employed. Although this bias can seem irrational, it's human nature to want what's in demand. That's why people tend to flock to the most popular movies, to buy the books on the best-seller rack at the bookstore, and to look twice at merchandise they've seen others use or wear.

By contrast, if you seem needy and easy to get, people will be less eager to hire you and more reluctant to make a handsome salary offer. As James Crowe, CEO of Level 3 Communications, says, "We want people who are in demand. They generally have to walk away from money. Then we will do what is appropriate [to get them]." Having a current job while you search for your next job enhances your image as a person who is in demand.

The job advancement approach laid out in this book is one you can follow *while* employed. As you'll see, it's not an approach to use only when you're in need of work; rather, it's an ongoing way to live your professional life and pursue your goals. And because you can't always control whether you get to stay in a job or go, it's important to prepare for the future and its inevitable uncertainties. One way to do this is by taking time to reflect on where you are and how you got there, both personally and professionally. This is called *setting your compass*, and you'll learn how to do it effectively in the next chapter. By setting your compass, you will be prepared for whatever comes your way.

3

SET YOUR COMPASS

"Nobody starts off to play an instrument with the idea of making money. You learn the guitar because you have this burning desire."

—Keith Richards, member of the Rolling Stones

Know Thyself

A business friend of mine—Garret—once played a mischievous trick on his employees. He pulled out their resumes from his file and made a few changes to them. He altered the employees' names, the names of the companies they'd worked for, and other details that would identify each employee. At the next staff meeting, Garret passed out copies of the resumes. "These are some folks we're thinking of hiring," he said. "What do you think?"

The results were startling. The team members didn't even recognize their own backgrounds. To make matters worse, they all agreed that they'd never hire any of these people!

"Know thyself" is a traditional bit of philosophical wisdom. It may sound simple, but as Garret's story illustrates, it's not so easy to do.

Knowing yourself is especially important when you're about to launch a new stage in your career. To help you know yourself, I recommend an exercise known as *setting your compass*. As you'll see, setting your compass is a process of taking inventory, of examining your past experiences in work, in school, and in life, as well as the interests, skills, knowledge, talents, dreams, goals, and preferences that these experiences reveal. The objective is to see your own background as others do, and to objectively review yourself—as Garret's team did, without realizing it—to see if and why you'd hire yourself.

Think of setting your compass as a way of mapping where you've been, so that you can more intelligently chart where you're going. You won't map out a step-by-step career plan that you expect to follow in detail. Given the changeable nature of the business world, that is unlikely to happen. Instead, setting your compass will help you determine where you *may* go and help you avoid places you'd clearly prefer to skip. Further, it will help you decide on a long-range objective, the larger goal toward which you'll always be steering, despite the inevitable zigs and zags along the way. This long-range objective will be your *Career Mission*.

Why is it important to set your compass and establish a career mission? There are many reasons.

First, knowing yourself is a crucial first requirement for effectively presenting and selling yourself. One recruiter I know remarks, "In interviewing candidates, I find that 60 to 70 percent of the people I meet try to present themselves as who they think they should be instead of who they really are." The problem with this approach is that the difference, although invisible to the candidate, is often apparent to the interviewer, which makes the candidate appear insecure, and perhaps even a little dishonest. Ironically, this form of deception is generally completely unnecessary. As my recruiter friend remarks, "Who they really are is usually just fine." Setting your compass makes it easier to know yourself and to present and sell yourself truthfully.

Setting your compass also serves to clarify the progression of your past career and illuminate the connections to the steps you hope to take in the future. If you're interested in changing fields—moving from a low-tech industry into a high-tech industry, for example—you may assume, as many people do, that your past experience is irrelevant. This is almost always false. Setting your compass will help you to uncover the skills you've developed in low-tech business that can be translated into success in the high-tech arena.

Setting your compass can enable you to define and sharpen your understanding of what you want out of work and life. Looking back at the most important, memorable, and meaningful activities in your past may illuminate inner drives and desires you never fully recognized before. If you've had, say, five different jobs in the past ten years, reviewing and reflecting on them to discover what they all have in common can help you better understand what makes you tick, making your ideal job that much easier to discover.

Those who've experienced the power of compass setting attest to its value. One entrepreneur I know, Ormann, has made it into a regular routine for himself. I asked Ormann how he'd gotten started with it. "The first time," he told me, "was during a hospital stay when I was in Vietnam. I had a lot of time to reflect." Today, Ormann asks all of his top employees to write ten- to fifteen-page histories of their own lives and careers, starting at age fourteen. Ormann believes that capturing on paper their personal and professional evolution leads his employees to powerful insights that are available in no other way.

Finally, setting your compass at the start of career- or job-changing time has the benefit of helping you to become more comfortable thinking and speaking about yourself, your achievements, your talents, and your goals. Jim McGovern, cofounder of Zissis McGovern, says it well: "It's packaging you. People take for granted what they've done. Even very accomplished people are unpracticed and often insecure talking about themselves." Yet the ability to talk about oneself is crucial to a successful job search. You can enhance that ability by setting your compass.

Step One: Creating Your Life Inventory

Once you're sold on the benefits of setting your compass, schedule a day on your calendar for working your way through the process—and I mean a whole day. With concentration and effort, it may not take a full day, but plan for it anyway. Don't allow other demands to distract you or steal time from the process—this is an important investment in yourself and your future, and you need to resist the temptation to postpone or "forget" it. Find a quiet place to work, where ringing telephones, family and friends, radio, TV, and e-mail won't intrude. If necessary, hole up in a secluded corner at your local public library or even a bench in the park.

You'll need writing supplies: your PC or a laptop, if you like to write on a computer, or a couple of legal pads and pens. Setting your compass starts with a written inventory of your skills, abilities, and experiences. You can write about something from first grade, from a first date, from a past job. The point is that it has to be memorable to you in some way. If it helped build your philosophy, your personality, or who you are in any way, it's important. This is your Life Inventory. The goal is to write down *everything* you can think of—to be all-inclusive. As you cast your mind back over your entire life, memories you may not have encountered for years are likely to rise to the surface. You may catch yourself saying, "Oh, *that's* not important," or "I don't need to write *that* down." That's wrong. Put it down! No matter how "trivial," "ancient," "silly," or "irrelevant" it may seem, put it down! Your Life Inventory is designed to capture the *totality* of the life and work experiences that have shaped you—not just the ready-for-prime-time highlights you usually focus on.

To get started on your Life Inventory, respond to the following items. As you'll see, the Life Inventory process is an extensive one. Of course, you don't have to write an essay on each point. A few sentences or a simple list of words or phrases in response to each item is sufficient. The key is to have an answer that makes sense and rings true for you.

LOOKING BACK

- List every paying job you've had, starting with your childhood and going up to the present day. Include part-time jobs, summer jobs, and temporary jobs as well as regular full-time jobs.

- List every volunteer, charitable, or educational work experience you've had, from childhood to the present day. Include community activities, internships, and other unpaid work.
- List all of your educational experiences to date. Include both full-time, formal schooling (from elementary school through college or graduate school) and part-time, on-the-job, after-school, distance learning, or adult-education programs in which you've participated.
- List the ten most important turning points in your career—for better or worse. These are moments when your life and work changed in a significant way, whether or not your realized it at the time.
- List your most significant past accomplishments at work. For each, explain in a few words what makes it important to you.
- List your most significant past accomplishments in unpaid or volunteer work. Again, explain what makes these important to you.
- List any awards, prizes, honors, publications, or other forms of recognition you've received for achievements in work or life.
- Which of your past paid jobs was the best? Why?
- Which of your past paid jobs was the worst? Why?
- Which of your past unpaid or volunteer work experiences was the best? Why?
- Which of your past unpaid or volunteer work experiences was the worst? Why?
- Which of your past supervisors, employers, or bosses was your favorite? Why?
- Which supervisor, employer, or boss was your least favorite? Why?
- List the people you've most enjoyed working with as colleagues or supervisors.
- What qualities, if any, do they have in common?
- List the people you've least enjoyed working with as colleagues or supervisors. What qualities, if any, do they have in common?

- For each job you've held in a for-profit company, ask yourself how you personally contributed to the company's profitability. Write down your most significant contributions to profit.
- What are the worst mistakes you've made in your career? What lessons did you learn from each one?
- What actions or inactions have you been criticized for? How did you react to the criticism? On reflection, was the criticism fair, unfair, or a little of both? Do you wish you'd reacted differently at the time? If so, how?

LOOKING AROUND

- List the things you most enjoy doing at your current job.
- List the things you are best at doing at your current job.
- List the things you most enjoy doing outside of work.
- List the things you are best at doing outside of work.
- List the personal qualities you are most proud of.
- List the things that most highly motivate or inspire you.
- List the things that most seriously demotivate or discourage you.

LOOKING AHEAD

- List all your current technical, professional, or managerial skills.
- List the skills you would most like to improve or obtain during the next five years.
- List what you consider to be your greatest personal strengths.
- List the personal qualities you would most like to improve or enhance during the next five years.
- Whom would you like to speak with about your career goals? Why? The answer could be a specific individual or a type of person. (For example, your response could be, "An expert on Web site design for nonprofit organizations.")

- What kind of place would you like to work in? Why? Describe your dream workplace in terms of size (big, medium-sized, small), history (legacy, start-up), and equity structure (individual- or family-owned, publicly traded, venture-backed).
- Where would you like to live and work? Why? Describe your dream geographic region and community.
- What values are most important to you in life? List your top ten values, then rank them from most to least important.
- What do you still want to achieve in your career? List the personal accomplishments that would mean the most to you at work.
- What do you still want to achieve in your life away from work? List these accomplishments.
- What do you want to maintain from your current work/career situation?
- What do you want to avoid in your future work?
- How would you define *success* for yourself? Write a brief description of what you would need to feel successful.
- List the most important rewards you want from work, including money. Then rank them in order from most to least important.

Step Two: Charting Your Career Trajectory

Once you've completed the Life Inventory process, it's time to begin analyzing your lists and other answers in search of meaningful patterns. The result should be a word picture that represents your Career Trajectory—the overall pattern that emerges when you study the things you've done, the interests you've developed, and the choices you've made. What you're looking for is concise yet meaningful answers to the following questions:

- What have you done?
- What are you doing?
- What do you want to do?

Ideally, your Career Trajectory will consist of a two- or three-sentence answer to each question based on your total life inventory. Encapsulating a lifetime's worth of experience in a few sentences isn't easy. It requires serious thought in search of the patterns, the common threads that connect the scores or hundreds of disparate, varied activities you've engaged in. The value of the Life Inventory you've created is that it brings out the variety and breadth of experiences you've had. Now it's time to simplify, to winnow, to uncover the hidden connections that make sense of it all.

If you look back over the pages of notes you've created and find them disjointed or confusing, don't been discouraged. The same kind of apparent confusion marks the lives of most people, including the most successful. Consider what Carly Fiorina, CEO of Hewlett Packard, said about her own life in a commencement address at MIT: "Journeys in life are far more random, far less orderly, than they seem at first glance. It's only when you stand back and see the whole journey in perspective, the paths chosen, the paths rejected, [that] a pattern emerges, a pattern that over time defines the journey of life. . . . I can see now that I started my professional journey on the day at age four when I announced to my parents and to the world, 'Mom, Dad, I want to be a fireman.' . . . I see now also that I began my path to become CEO on the day I decided to quit law school."

WHERE YOUR JOURNEY BEGAN

Where and how did *your* personal journey begin? Did your childhood career dream (fireman, ballerina, football star, astronaut) lead directly into your later work? Did you find new and different ways to express some of the same aspirations and values (heroism, beauty, strength, adventure) while following a very different career path? Or did you abandon your childhood dreams prematurely, leaving you with regret that you've never quite overcome? (If so, perhaps it's not too late to find a new way of realizing those dreams.)

Carly Fiorina now sees that quitting law school was the first step on her path toward becoming a high-tech CEO. What were the significant turning points in your life? How did your choice of college major, first "real" job, partner, place to live, or career shift shape your later experiences? Look back at the paths you said "No" to—the job offer you turned down, the school your parents urged you to attend that you rejected, the relationships you broke off, the promotion you didn't accept. How might your life have been different if you had said "Yes"? Do you feel any regrets, or only a sense of relief?

THE CHOICES YOU'VE MADE

Consider the dozens of life and career choices you've made over the years. Try to remember how and why you made those choices. What were you expecting at the time? How did the reality match up with your expectations? Did the choice of school, career, job, or spouse work out better than you dreamed—or far worse? What do you know now that you wish you'd known at the time? How might you have made a more informed and wiser decision? What can you do to improve your decision-making the next time you're faced with a life-altering choice?

THE PATTERNS THAT EMERGE

Finally, look for patterns—hidden similarities that underlie the choices you've made. On reflection, you might discover that you've repeatedly chosen the easier of two career paths—or the harder. (Neither of these patterns is necessarily "right" or "wrong"—what's important is that you recognize your own tendencies and make conscious choices about whether and how to change them.)

Perhaps you've repeatedly taken on job assignments and challenges that were a little beyond your ability to handle—or perhaps you've had a tendency to shy away from projects that you feared were too difficult. Perhaps you've tended to become bored after one or two years doing the same kind of work—or perhaps you've settled easily and comfortably into a routine and avoided change thereafter.

Maybe there's a pattern in the kinds of jobs you've loved and hated. Maybe you like working in an atmosphere of crisis, where time and money pressures demand heroic efforts that are exhausting but exciting. Or maybe you're just the opposite: You prefer a workplace that's calm, methodical, and orderly, with well-regulated processes and plenty of resources.

Maybe you love working closely with other people, responding to their needs and problems with sympathy and good humor. Or maybe you prefer being left alone, working on your own projects according to your own time schedule. Again, there's no "right" and "wrong" preference, but it's crucial to understand your own inner workings and how they've expressed themselves in your Career Trajectory. The better you know yourself, the better you can make choices that will lead you to a satisfying and rewarding future.

WHAT IT ALL MEANS

It's important to be able to recite your accomplishments, talents, and strengths. But that's not enough. It's just as important to recognize the psychological and emotional wellsprings that those outward expressions reflect. As Steven Aldrich, president of QuickenInsurance, recommends, "Focus on the *why* as much as the *what.* I interviewed someone the other day and I could not understand what motivated the person or what he had learned, just what he had accomplished. That's not going to cut it."

Step Three: Developing Your Skills Inventory

Setting your compass isn't only about where you've been, but where you're going. And where you're going depends on your value to a company. To develop a realistic sense of what that value might be, you have to fully review what you've done in a systematic fashion. Then take what you've done and translate it into terms that express your potential value to a future employer.

The first important step in doing this is to develop a Skills Inventory. This is a comprehensive list of all the abilities you've developed and used throughout your life, both on the job and off the job. It will probably include several kinds of skills: technical skills (knowledge of Excel, ability to program in HTML, skill in graphic design, for example), managerial skills (such as ability to administer a budget, skill at motivating people, team-building prowess), industry knowledge (understanding of current trends in telecommunications, for example), and personal skills (such as ability to thrive under pressure, willingness to work with all kinds of people, skill at defusing tense situations). Think in terms of both "hard" skills (those that are quantifiable and often taught in formal educational and training programs) and "soft" skills (those that are harder to define and developed through life and work experience). Both are important and valuable to employers.

To create your Skills Inventory, review all the items in your Life Inventory. Think about the jobs you've held, the unpaid work you've done, the achievements you've had, and the experiences you've enjoyed. Chances are good that each of these items can suggest one or more skills that belong in your inventory. Try to judge yourself objectively, as an outsider (or a potential employer) might, being neither too easy nor too hard on yourself.

Many people, especially career changers, fall into the trap of overlooking or undervaluing their past experiences and the skills they represent. You may catch yourself thinking, "Oh, now that I'm moving into the high-tech arena, no one cares about the things I accomplished when I was working in real estate" (or retailing, or at the post office, or whatever). This thinking is flawed. Remember, the high-technology arena is so new that it boasts very few longtime veterans. *Most* of the people working in high tech have spent the majority of their careers in traditional industries. Furthermore, a lot of traditional roles and skills are transferable to the high-tech industry. If you have innate intelligence, enthusiasm, ability to work effectively with other people, and a strong drive for accomplishment, those qualities are just as important as specific prior experience in high tech—sometimes more so.

This truth is becoming more and more widely accepted today. After all, a major cause of the dot-com meltdown of 2000 and 2001 was the fact that many high-tech start-ups lacked traditional business talent and (somewhat arrogantly) assumed that it was no longer necessary. In the wake of the downfall of many companies that pinned their hopes on technical prowess alone, smart high-tech businesses are scrambling to bolster their strengths in such traditional areas as sales and marketing, finance, human resource management, production, and logistics.

Some of today's leading high-tech entrepreneurs owe their success to their strong backgrounds in traditional business. Michael Dell, founder and chairman of Dell Computer, learned to connect with customers during his early years at Procter & Gamble. Bob Hornung, the founder of Onlink.com, told me, "When I started my business, I didn't even own my own PC. I had never been in software. But because I'm a great people picker, I created the best business around." Onlink.com had generated $2 million in sales the year before it was purchased by Seibel Systems (for $760 million!).

Of course, experience and familiarity with technology are big pluses when seeking a career in high tech. However, your Skills Inventory should include *all* your business and personal talents, not just those that are focused on technology. Also, remember that neither experience nor skills can necessarily be measured in years on the job. Some people achieve, learn, and grow enormously during twelve months with a company; others may stagnate and accomplish little over a decade or more. So don't automatically write off skills or expertise you may have developed during a brief three- or six-month stint. They may be enough to put you over the top in competition with other candidates for a job you'd love. As David Selden, CEO of ProBATE Software, puts it, "People assume that years of experience qualifies you for a job. It's what you've done with that experience that determines whether or not you are qualified for a position."

Reality Check

The ultimate value of setting your compass is that it enables you to draw realistic connections between you and the job of your dream. In effect, this is a four-part task that includes:

1. Analyzing your skills and experience as a prospective employee
2. Analyzing the skills and experience sought by the marketplace of employers
3 Looking for matches between what you have to offer and what employers want
4. Taking action about any gaps that exist between what you have and what employers want

It's a tactical and strategic approach that works for your overall career mission by defining where you are, where you need to be, and what you must do to get there. It's also a way of managing your career in the same way savvy businesspeople manage entire companies. Michael Doyle, a senior brand director at Immedient, uses the same basic process in new-business development. He says, "To move up the food chain, I ask, 'What kind of [new] client do I want to get?' and 'What do I have to do to get there?' "

Although it seems natural to take this approach in building a business or developing new markets for products, we often fail to apply the same technique in personal development. One search consultant told me, "You have to view yourself as a product looking for a need. Pretend you are a box of Tide. How would you go about marketing that box of Tide? What makes you better than the other products on the shelf?"

What if, in your private, honest evaluation, you don't have a lot of past accomplishments or present-day skills to point to? Now is the time to face that issue and begin to deal with it. Setting your compass should be a reality check. And, in business, nothing speaks louder than quantifiable on-the-job results. One recruiter told me, "I just interviewed a woman with a twenty-year career—but she didn't have any results." If you look back over your work history and can point to very little in the way of measurable accomplishments—especially those that contributed to your employer's bottom line—then you need to ask yourself why.

Would your past supervisors or colleagues use any of the following phrases to describe you? If so, then you probably have work to do that's more important than job hunting.

- "Lacks discipline"
- "Can't execute and pull things together"
- "A real plodder"
- "Constantly churns from company to company"
- "Treats people as if they are expendable"
- "Shamelessly greedy"
- "Spends more time worrying about who gets credit than getting the job done"
- "An incessant complainer"
- "Uses company time for private projects"
- "Bad-mouths other people"
- "Calls in sick every other Monday"

Are your ears burning? If so, your career is heading for trouble. A reputation that includes any of these attributes does not make you a flawless contender, a "water-walker," the kind of person who exceeds everyone's expectations.

By contrast, the person with an impeccable reputation as a hard worker, a creative thinker, a caring colleague, and a positive presence in the office is likely to be much in demand—even when he or she isn't looking for a job change. Janice Aune, CEO of Onvoy, describes what it's like to be a sought-after superstar: "When they found out I was looking for a job, I heard from every previous boss I've ever had wanting me to come back and work for them or consult to them. I still get twenty or thirty calls a day from people who've worked for me, saying, 'If you're going to do something new, let me know—I want to do it with you.' "

Of course, not everyone can be a superstar, deluged with calls from headhunters and bosses, but if you've been working in your field for a few years and receive calls from recruiters rarely or never, the market may be sending you a message—one you'd be wise to heed. (Chapter 4 provides a number of specific suggestions about steps you can take to enhance your value in the high-tech career marketplace.)

JANA'S STORY

Jana Weaver was formerly a manager of Web marketing with Quantum. When she wanted to change her career direction, she began with a compass-setting process. Here, in her own words, is Jana's story of how she did it and what happened next.

> After working for a large, well-established Fortune 500 company for seven years, you would think that making the transition to a start-up would be hard. Well, it was. But making the transition to any other job would be hard, too.
>
> I'd known that I needed to leave Quantum for a long time (more than a year), but the timing never seemed right. Finally, I realized that I was at a turning point. I either needed to leave right away, or be stuck there for at least three or four more years. Why? Because I knew I wanted to have a baby, preferably in about two more years. I would not feel comfortable going on maternity leave when I've been with a company for less than a year, and two years would be best.
>
> So I was on a train, and I had an opportunity to switch tracks. If I didn't switch tracks then, it was going to be a long time before the opportunity came up again. So I did it.
>
> The process of looking for a new job, and narrowing down exactly what I wanted to do, was a long and arduous one. I started really thinking about what I wanted to be doing, and for what type of company, about three months before I ever sent out a resume. I wanted to make sure that I made a sound decision when I finally did leave, and I didn't want to leave just for the sake of leaving. (That isn't always a bad idea, but you never want to go from the frying pan into the fire!)
>
> I made many lists. Lists of characteristics my ideal company and ideal job would have. I listed things such as "an open, friendly environment" and "ability to continue my personal growth." These ranked higher for me than "top salary."
>
> I found I was torn between moving up in the field I'd already been working in for the past five years and making more of a lateral move within the marketing field to a different specialty. On the one hand, I knew Web development and management really well. On the other, I knew I'd have trouble getting much higher within a market-

ing organization (becoming a VP of marketing, for instance) without some more solid product marketing experience. In the end, I decided to explore both.

When I was ready, I went to all the popular job search sites and submitted my resume for any jobs that looked as if they met my criteria. The first night, I sent out twenty-eight resumes. Within a week, I had sent out more than fifty. I also sent my resume out through my network of former colleagues and friends. I spoke with probably twenty-five companies on the phone, had in-person interviews with twelve, second round interviews with five, and offers from three.

One of the trickiest parts of this whole process is timing. You don't want to get one offer and have to make up your mind without the benefit of having at least one more. Ideally, you will get at least three offers, all around the same time. Unfortunately for me, there was a two-week time lag between the time I got my first and second offer, and the third offer was only a verbal one. (They couldn't officially extend an offer until they got their second round of funding.)

Of the two "real" offers I received, one was from a pre-IPO company to do the same job I'd already been doing for many years. The other was from a public company doing a totally new job, but in a field I was already knowledgeable in. I didn't set out to work *only* for a start-up, but it did turn out that more than half the companies I met with were pre-IPO.

I found that there is a big difference when evaluating offers from start-ups versus public companies. Most likely, the start-ups will offer a much larger number of stock options, and at a very reduced price. However, those options are worth *nothing* at the time, and they may never be worth anything. On the other hand, the public companies usually have more programs in place, such as bonuses, employee stock purchase programs, and many perks that the smaller companies can't afford. I thought about all these things when evaluating my choices.

However, in the end, I made the decision based mainly on the people I was going to work with. The public company I interviewed with had just lost the director I would have been reporting to. Additionally, there were some bizarre things that occurred during my interviews with them that made me question just how together they were.

4

POLISHING THE PRODUCT

"I couldn't wait for success, so I went ahead without it."

—JONATHAN WINTERS, comedian

What Makes a Superstar?

The world of technology is exciting. Exciting not just because of the way it's changing our lives, but also because of the potential personal profit. As hundreds of thousands of people have already discovered, it's possible to make a great living, and even amass considerable wealth, through a career in the high-tech universe. However, that kind of success and excitement doesn't just happen. *You must make it happen.* And you make it happen by offering a high-tech employer something of unique value—by being indispensable and irreplaceable.

It only makes sense. After all, if *you* are the product, isn't it logical to make the product the best it can possibly be before you begin shopping it around?

In Chapter 3, you took time to list the accomplishments of your life so far. Now think about them as a whole. If you were to rate them on a scale from 1 to 10 (with 10 being the highest), where would they rate?

Now think about the role you play in your current job. How do you think your boss would rate your contributions to the company's bottom line? Finally, look into the future. Think about the job you'd like to have one year from now. How do your skills rate as measured against the demands of that job?

Suppose you can honestly rate yourself a 9 or a 10 in your current work. That means your job performance is outstanding. The quality of your work is high; you're hard working and serious about the job; you apply intelligence and imagination to every task you're assigned. All of that is important and impressive. But it's still not enough. In today's business climate, being good at what you do is *necessary,* but not *sufficient.* Many people are highly competent, smart, and productive. Relatively few are also memorable, impressive, credible, genuine, trusted, and liked. These additional qualities are what lift a handful of skilled performers into the realm of superstardom.

Of course, some people enter adulthood with more of these qualities than others. Thanks to some combination of genetics, family upbringing, education, and early work experiences, they know how to think, act, and communicate like winners from their first day on the job. But it's also true that virtually anyone can increase their share of these vital qualities. All it takes is a healthy dose of self-awareness and a willingness to work hard at changing and growing, both at work and in life.

In this chapter, you'll learn in detail three of the best ways of enhancing your own star qualities—especially as related to the world of high technology. These three ways are:

1. Learn all you can about technology.
2. Operate according to a new-economy mentality.
3. Live the best of the old economy values.

Let's consider each of these.

Learn All You Can About Technology

I'm a great admirer of Ernie Howell, the retired president of WPM Systems. Every semester, he signs up for a few more university classes in a wide range of disparate fields. Why? "I've got to learn everything I can before I drop dead," Ernie explains. It sounds like a funny way to put it, until you realize that Ernie is seventy-seven years old!

You and I could do a lot worse than to imitate Ernie Howell. In fact, in today's career marketplace, doing what Ernie does isn't just a nice idea—it's a necessity. Today there's just so much more to know and to get better at than ever before. In addition to traditional business subjects such as strategy, persuasion, marketing, finance, leadership, vision, and people skills, there's a never-ending stream of new technologies to learn about.

To be successful in the high-tech world, you don't have to be a techie, but you do need to have a strong technical curiosity. Work on knowing all you can about the technologies you and your current company are using now as well as those you'll be using in your next job. It doesn't take extraordinary intelligence—just a willingness to work at learning. Software engineer Steven Schimmel says it this way: "The barriers to success in any company stem from a lack of dedication to learning. Other than that, the barriers to entry into high-tech industry at any level are relatively small."

You'll find that the incredible speed at which technologies change is both a curse and a blessing. It's a curse in that the need for learning never stops. If you pick up a trade magazine in your industry and realize that you have no idea what the articles are about, you'd better get up to speed—fast. In less than six months, your existing knowledge may be totally obsolete.

But the (sort of) good news is that if you've missed the latest technology round, you may be able to get in on the next one. Today, *any* individual aspect of technology is likely to have a short life cycle. A recent Hewlett-Packard (HP) study tracked the revenues earned from newly introduced products. In Year One (the launch year), revenues were generally small, but growing rapidly. In Year Two (the year of widespread adoption), revenues were high. And by Year Three, *revenues were already in decline.* In fact, HP found that the demands of competition and the opportunities offered by technological advances were dictating an almost complete turnover of their product line every four years.

Thus, as Kevin King, president and chief technologist at Precision Solutions, warns, "Your job will get obliterated if you don't keep up. But," he adds, "you can ramp up in eighteen months and catch the next wave if you missed the last one."

Among other implications, the current speed of innovation means that the career advantage enjoyed by certified, dyed-in-the-wool techies may be much less than you assume. Markus Schweig, a manager at Microsoft, observes, "Even if you grew up with high tech and are totally receptive to it, the technology you learned in college is now learned by high school students. And technology has changed a hundred times since you got out of school." The pace of change, in effect, levels the playing field every two or three years, forcing everyone to go back to school (literally or figuratively) and start fresh. So if you feel as though you're a few laps behind as you start the race, don't worry—if you work hard, you'll be ready to start the next race virtually on a par with your techie competitors.

THE VALUE OF CONSTANT LEARNING

Another corollary of the speed of new developments is the fact that new technology pays much better than old. Everyone recognizes how this works in the world of products: The introduction of Pentium-level chips for PCs instantly drove down the price of the old 486-based computers. The same thing happens in the job market. As I write, the C++ programming language is almost two decades old, while Java is only about five years old. Consequently, Java programmers earn much more than C++ programmers, because there are fewer of them and the demand is higher.

Software engineer Kevin Werner has absorbed this lesson. "Those who keep learning will be successful. Anyone who thinks the changes will start to slow down and society will 'settle in' on the Information Age is kidding himself. . . . I am a Java software engineer, but in five years I may be something else, because soon something else may be taking the place of

Java. In technology, it is people who see where the technology is going who make the most money. . . . I have friends who started programming Java before it became so popular, because they saw the potential. They are now making $100,000 to $500,000 a year."

Perhaps it's understandable that people who are constantly learning have greater economic value. After all, those who make a habit of learning:

- Can do more things more quickly and effectively
- Are able to make the most of their productive abilities
- Are more flexible, versatile, and forward-looking
- Respond promptly to shifts in customer needs and preferences
- Tend to be more imaginative, creative, and innovative
- Have more to offer their teammates and colleagues

Thus, in the high-tech world, learning isn't just nice to do; it's a *must do* for anyone seeking lasting success.

SEIZE LEARNING OPPORTUNITIES

Here are a few specific tips about how to make the most of the learning opportunities available to you.

• **Learn to pick winners.** Of course, not every new development is worth mastering. Just keeping up with the flood of new discoveries, inventions, and theories is close to a full-time job. You need to be selective about which new technologies you'll devote significant time to exploring. In picking topics that deserve your serious attention, consider the following questions.

- Is this something I want to learn? Is it intriguing, interesting, challenging, exciting, and truly new?
- Is this something I need to learn? Does it relate to my current job or to the next job I hope to have? Is this a technology that is already or will soon be in use at my company or at competing firms?

- Is there a clear and present need for this technology? Is it possible to identify a significant market of individuals or businesses that have a definable way to use and benefit from it?
- Is this technology likely to be profitable in the foreseeable future? Is it likely that the identifiable marketplace will be willing to pay enough for this technology to cover the costs of its development and production as well as a reasonable profit?

If you answered "Yes" to three or more of the above questions, you should probably put this new technology on your To-Learn list.

• **Learn a little about everything.** Your regular media menu should include magazines, newspapers, columnists, conferences, and conventions that cover a broad spectrum of technology topics and related fields. Don't focus too narrowly on your current job description or your company's latest strategy; after all, both are likely to change sometime in the next year. For example, if you're involved in broadband telecommunications, you need to know at least a little about the convergence of various entertainment and information media, about cable, satellite, and wireless technologies, and about the latest developments in laptops, handhelds, and other mobile computing gadgets. Similarly, if you're with a genetic technology firm, you should understand the implications of nanotechnology, the latest developments in pharmaceutical research, and other "neighboring" fields that are sure to affect your company's work within the next few years.

• **Learn everything about something.** In addition to reading and learning broadly, learn one field in depth. Keith Johnson, vice president of ThinkArena, puts it this way: "Pick an area of technology and know it inside out and upside down. Know the competitors, the business space, the market size." Delving deeply into one area—learning everything there is to know about how the technology interacts with economic, social, political, legal, and demographic factors to create a particular kind of business opportunity—makes you a true expert and, therefore, a valuable person to have around.

• **Take advantage of every learning opportunity.** There are plenty of places to acquire tech knowledge: college and university courses, lectures, seminars, self-study, professional societies, conventions, conferences, and in-house training programs. Costs will vary widely. An evening course offered by a private training company or the continuing education department of a university might cost $500. A week-long course in Java programming offered by a licensed company like Allaire Corporation or one of their training partners may cost between $3,000 and $5,000. By contrast, there are hundreds of excellent reference books suitable for self-study purposes at prices in the $50 range. There is even federal aid money available to displaced workers for job retraining. Experiment with some or all of these, depending on your needs, interests, schedule, and budget, as well as the support your employer provides.

Some jobs offer learning opportunities that are fully comparable to classroom programs. Consider applying for internships and consulting assignments (using skills you already have in new settings or businesses you'd like to learn about), as well as volunteer tasks in not-for-profit organizations. Be open to learning anywhere, from anyone, with or without the guidance of a teacher.

• **Ask your employer for help.** The American Society for Training and Development recommends that every company set aside about 4 percent of its total payroll for its employees' continuing education, and more and more smart businesses are following this advice. In fact, the availability of educational help has become a major factor in determining how successful companies are at attracting and keeping good workers. Statistics show that companies who *don't* invest in developing their people are more than 300 percent more likely to lose them.

If your company is investing in new technology, it's likely they'll want to pay for additional training in making the most of the new systems. It's simply a competitive necessity. As Dale Telford, founder and CEO of bITssol, Ltd., says, "With new technology changing so fast, anyone who sits on what they've built or invested in will be lost. . . . Since I've come into this industry, we've evolved from XTs that ran at 8MHz to third-generation

Pentiums running up to 1GHz. . . . It's too risky to try and push the extra year out of the systems you invest in. I could tell you a dozen or more stories where I've seen it destroy a company." In some cases, compliance with government regulations and industry standards actually requires continuous employee training.

Is it wrong to allow your present company to pay for your training if you are considering changing jobs or industries? It depends on several factors, including how long you expect to stay at your current job, whether or not you are considering a move to a direct competitor, and whether you'll be able to put what you learn to immediate use for the benefit of your present employer. Any smart employer knows that there is a good chance that the money they invest in training their people may "walk out the door" in the future. The best companies continue to invest in education despite this risk, trusting that they'll benefit in the long run from contributing to higher standards of knowledge for their entire industry. One executive with the right idea comments, "When people worked for me, I'd pay for training to upgrade their abilities. Then, if they got hired elsewhere because of their improved skills, I'd say 'Go for it.' I was glad any time anyone left to better herself. In fact, I'd say, 'Leave, so I can hire you back [later] at more money.'"

- **Make learning your own responsibility.** Although many employers willingly support their workers' continuing education, don't wait around for your company to train or develop you. At some firms—especially at start-ups and other companies where time and money are in short supply—the onus falls entirely on the employee. Steve Cornforth, director of human resources at TrueSpectra, speaks for many companies when he says, "We don't develop people—we don't have the time. We need talented people now. We don't have the time to send them to a course. They have to move up on their own." This means that you should take all the help your company will give you, but seek out your own opportunities as well.

- **Have the courage to learn.** Given the power and importance of learning, you might assume that everyone in the high-tech arena would pursue it aggressively. Yet many high-tech workers don't. Why not? "Most people prefer to stay where they are," says Jeff Joerres, CEO of Manpower. In a sense, it's understandable. Learning means changing, and changing always involves taking risks—exchanging the known for the unknown, the proven for the untried, the comfortable for the risky. No wonder most people quietly avoid the challenge of learning. But you can't be "most people" and position yourself for the leadership job, the cutting-edge position, or the big money. It takes passion and commitment to do that.

Operate According to a New-Economy Mentality

Several of the CEOs I interviewed for this book made the same point. "Don't let your readers become obsessed with technical qualifications. They can always get those. What is important today is to have the *new-economy* mentality." What does this mean? Where do you get it? And most important, how can you let yourself be guided by it?

Unfortunately, the dot-com flameout has given the words *new economy* a bit of a bad aftertaste. Let me begin by telling you what the new-economy mentality is *not.* It's not the get-rich-quick business model of "Develop an idea, attract investors, rush to market, cash out, and profits be damned" of the late 1990s. Instead, the new-economy mentality is characterized by the following working principles.

- Become productive *fast.*
- Think before you leap.
- Admit your mistakes.
- Add value to the team.
- Welcome change.
- Take risks.
- Make do with what you have.

Let's consider each of these principles in some detail.

BECOME PRODUCTIVE *FAST*

In the new economy, you can't wait around for things to happen. Nor can you wait until everything—or anything—is absolutely perfect. Nor can you wait until you "get a feel" for things. You have to become productive right away—your first day on the job.

Speed is of the essence in the high-tech arena. Try to come up with a good idea every day—stuff that will make a positive impact on your company's future. But coming up with new ideas is the easy part. Getting them to work is hard. Getting others to accept them is even harder. And it's even harder to innovate when you're a new member of an established group. Nonetheless, this is your assignment: to be the one who shortens the distance between the great idea and its successful execution.

A former COO of Amazon.com described his first two weeks at the company this way: "It was like someone turned my treadmill up to the fastest possible speed."

THINK BEFORE YOU LEAP

Speed doesn't mean acting without consideration for the future. It does mean timing the market. Move too fast and you'll pile up losses; move too slowly and you'll be left on the sidelines. Speed means finding the *right* speed for the rapidly shifting business world.

Before acting quickly, think your decision all the way through. I'm not suggesting old-economy deliberation, ponderous and slow. I'm suggesting a little Monday-morning quarterbacking *before* Monday, while you still have time to find the open receiver. Yes, you need to be out and about, talking, asking, listening, sharing. But you also need time in the quiet of your office or home to sit and think before acting.

Only through careful thought can you "know when to hold 'em and know when to fold 'em," in the words of the old country song. Both choices are important and valuable in the right situation. As Larry Feld, president of Human Performance Strategies, puts it, "Don't confuse the fold with failure or defeat. Rather, it's a reallocation and redirection of precious resources."

ADMIT YOUR MISTAKES

Moving quickly can cause some mistakes. Don't fear or hide them. And remember, if you've had some early wins, you'll be forgiven more easily for a few early mistakes.

When you do make mistakes, admit them; don't hide them under the rug, hoping no one will notice. People *will* notice, and they'll also notice your attempted cover-up. Act openly to reduce the impact of a mistake, learn from it, and be prepared to state clearly to everybody exactly what you've learned.

Finally, forgive yourself. Not only is this a necessary component of mental health, but it makes it a lot easier to forgive others for their mistakes when they make them—and they will.

ADD VALUE TO THE TEAM

The best job advancement technique is doing good teamwork, instead of just being out for yourself. This doesn't mean that your goal should be to blend into the crowd. Good work means solving the toughest problems. Jump on those—go out of your way to stand up and volunteer for those opportunities. Just do it in a team spirit of service, not one of self-aggrandizement.

Are you really smart, really talented? If so, great. Let people discover that on their own. "You may be a bright light in technology, but you still can't act like you're smarter than everyone else is," warns Ron Dobson, president of ADP Canada. "It's a deadly sin." In fact, the person who develops the habit of thinking *I* instead of *We* is apt to end up on the outside looking in. Charlie Wallace, a managing partner at the Heidrick & Struggles recruiting firm, says, "I listen very carefully to see what the person says about her team. I want to hear whether she acknowledges the contributions of others or dismisses them."

WELCOME CHANGE

Technology means finding better ways of doing things, which means *change*. Thus, the biggest perk of technology is freedom to be original. Take advantage of that freedom. Be different. Be unique, inventive, imaginative. Will this take getting used to? Of course. Will it be worth it? Definitely.

"When my business is stagnant for three weeks, I get worried," says Keith Johnson, vice president of ThinkArena. All those who've achieved success in the technology arena echo his sentiments.

TAKE RISKS

Leaders from the tech sector talk a lot about the need to take risks—risks in new technology, risks in timing, risks in investing huge money with no guarantees. These are all important for success, but perhaps the greatest risk arises out of one-on-one relationships with other people.

In the end, risk-taking is a profoundly personal issue. To take risks you need to feel secure, to believe in yourself. Being uncertain, apprehensive, tentative won't cut it. People don't follow and support those who appear insecure. So swallow your fear, take a deep breath, put a smile on your face, and act confident.

You weren't born with self-doubt; you learned it. So unlearn it. Teach yourself not to worry, stress out, procrastinate, or make excuses. Be pleasantly assertive as you use your guts, pluck, unbridled spirit, and controlled abandon. You're only as good as your courage, so first do the thing you fear most. You never know how it will change you. Just look forward to the unknown.

MAKE DO WITH WHAT YOU HAVE

The smaller the organization, the younger the management, the newer the business model, the riskier the idea, and the scantier the resources, the more you have to make do with what you have. In most high-tech companies today, you're either great or you're out of here. You'll be considered extra valuable if you make do and make the company better than it was when you arrived.

Of course, in the high-tech era, "making do" isn't what it used to be. The power of technology is available at shockingly low cost. With a $700 computer and Internet access costing $15 a month, you can buy and sell products and services anywhere in the world. And technology makes it easy to piggyback on other successful businesses. As one executive explains, "Take eBay, for example. Not only does eBay make money, but many entrepreneurs run their own Internet businesses by using eBay as the advertising and communications vehicle between buyer and seller. So for less than $1,000 you can have a company with an international market."

If you've been working at a legacy company or any large firm, you may find the lean-and-mean atmosphere at many high-tech start-ups disconcerting at first. There may not be support staff, systems, procedures, equipment, or budgets like those you're used to. The modus operandi at many start-ups is "Need something done? Do it! Don't know how? Learn it! No time or money? Make do!"

Most people find that once they become accustomed to the new rhythm, making do is exhilarating and fun.

Live the Best of the Old-Economy Values

Again, some definitions are in order. When I refer to old-economy values, I don't mean the slow-moving, ponderous, bureaucratic business approach that still plagues some legacy companies (including a few in the high-tech arena). I mean the best qualities found in the traditional businesses that built and shaped the economy we enjoy today. Over the past century, legions of smart, dedicated, hard working "grays" not completely unlike today's smart, dedicated, hard-working "punks" built profitable companies with their *interpersonal skills, positive attitudes,* and *ethical behavior.* These values will never go out of style.

Let's take a closer look at each of these three.

INTERPERSONAL SKILLS

Technology matters, but people matter more. People skills were the number-one skill in the old economy. In the new economy, they're even *more* important. People today have too many job choices to have to put up with jerks at work.

The extraordinary time pressures and high-risk environment of high-tech business put a premium on people skills. "Being motivated and liked can make up for a lack of skills," says Phil Duong, chief technology officer at Iconixx. "We work together twenty hours a day, four to five days a week. Things need to get done in four weeks, not four years." In that kind of atmosphere, prima donnas are not welcome. Hard-working team players who know how to build and maintain relationships are.

Ross Brown, vice president for human resources at Analog Devices, has this take on the subject: "Socio-techno is so complex, and you *must* operate in teams—even teams spread around the world. The soft stuff [like interpersonal skill] is hard, but in the final analysis, the soft stuff is what makes the difference." Thus, if you have good people skills, you'll be forgiven a lot if, when the dust has settled after a problem, people can still say about you, "Yeah, but I like her."

Of course, the term *interpersonal skills* is broad. What are some of the specific talents that make up this category of skills?

One is the ability to speak in a language your audience can understand. It's a vitally important skill that can only be developed through sensitivity to other people's needs, interests, and knowledge, and a lot of practice. If you can learn to make the complex issues simple, and serve as a communication liaison between the tech and the nontech worlds, you'll be an exceptionally valuable person to have around.

Use stories, analogies, and anecdotes when you can. Stories get people involved in your thought processes and paint pictures people can relate to. They humanize you, make you more colorful, and set you apart from the crowds of businesspeople who only speak corporatese. Just make certain that the stories you tell are true, appropriate, well told, concise, and fresh.

Learn to write concise and thoughtful e-mails, letters, memos, and notes. Start by pre-thinking your objectives and the points you want to make, then write from the heart. If time permits, put the document aside to revise later. And when revising your writing, trying reading it out loud. Does it sound like you? If not, revise it again.

When you've achieved a success, share the glory. Give others credit even if they didn't do all that much. It ends up elevating you as a team player. Maintain other people's self-esteem. Praise them publicly for what they do right; correct them in private for what they do wrong. Give your teammates opportunities to shoulder responsibility, to make decisions, to express their creativity. Don't berate people for their mistakes—instead, encourage them to learn from them and move on.

If you do all this, people will put more effort behind you and your projects. But if you strip people of their self-esteem through criticism, sarcasm, abuse, or neglect, you will find yourself on your own.

Maintain your own self-esteem, too. Don't undervalue yourself or let others do so. Learn to walk the fine line between excessive modesty and showing off. Speak up when you have something to say; respond with strength and dignity when you're disrespected or undermined.

Be willing to sell yourself and your ideas, and learn how to do it effectively. Learn to sell to your boss, your peers, your subordinates, and your company's customers. And learn to influence people who don't report to you through the intelligence, clarity, and accuracy of what you say. One headhunter says, "A person making $100,000 can make $10 million by helping people make decisions and influencing people who don't report to them." Is this an exaggeration? Not much of one.

A POSITIVE ATTITUDE

In life there is so little we have complete control over. The one area in which we have total control is our attitude. Fortunately, this is a powerful tool. In the long run, your success will be determined more by your mental attitude than by your mental abilities.

The key is being consistently positive, even when you don't feel like it, and reflecting that positive attitude in your posture, your voice, your handshake, and the decisions you make. It means not giving up even when some others think you should. It's following through on your personal commitment to results regardless of what you have to endure along the way. It means getting up when you're knocked down. Anything less is self-indulgent.

Being positive also means being cheerful, congenial, friendly. Don't worry, it won't take away from the seriousness and significance of your work. This doesn't mean you have to act like a stand-up comedienne, a gagster, or a prankster. And it definitely doesn't mean you should use humor to wound others. Sarcasm and ethnic, religious, or sexual slurs are never acceptable. The goal is to learn to express positive energy through a smile, by being willing to laugh at your own foibles, and by not taking yourself too seriously.

ETHICAL BEHAVIOR

No matter how smart, quick, and gutsy you may be, if you aren't honest, genuine, fair, and frank, you aren't going to succeed in either the new or old economies. At least not in the long run. The truth is that people can and do get away with being dishonest—but not for long.

In my book *How to Think Like a CEO*, I wrote about the ethics of near-perfect CEOs. You may not be a CEO or even an aspiring CEO, but the rules I learned in researching that book still apply. They include the following.

- Don't intentionally mislead or misrepresent.
- Don't straddle the line.
- Don't break promises or go back on your word.
- Don't "put a spin" on information.
- Do be completely honest with yourself.
- Do remember that regardless of how carefully you disguise the truth, someone will always find out.
- Do remember that your distrust of others will often justify their distrust of you.
- Do work to understand misunderstandings.

- Do realize that one person's honesty could be another person's dishonesty.
- Do work to bring uncomfortable issues into the open.
- Don't exaggerate. Say it like it is.

If you think of the world of new technology as being a world of hype, corner-cutting, flimflam, promotional puffery, phony deadlines, broken promises, and other chicanery, you're not wrong. High tech has indeed been plagued by these kinds of ethical shenanigans. It's the inevitable consequence of a gold rush mentality. One of the benefits of today's cooling-down period in the high-tech marketplace has been a return toward sanity and a reassertion of ethical standards. Don't jump into the high-tech business if you think it's a refuge for shady behavior. Those in the industry who are concerned with building genuine, long-term success are determined to do it ethically—and if you aren't, you're not likely to succeed.

The Merging of Old and New

"As the old-economy workers adopt the best of the new economy, and vice versa," says Ken Hayashi, CPA, "I believe the distinction will probably disappear." The blending Hayashi speaks about is already happening. There are already many companies that are almost impossible to define as either old- or new-economy businesses. In fact, they're artful blends of both. Consider GE, a classic industrial conglomerate that is rapidly transforming itself into an Internet-driven organization. Or IBM, an old-line maker of computing machinery that is retooling itself as a software and services company for the new millennium. Or consider AOL Time Warner, the product of a literal merger between old and new communications media.

As the old and new economies increasingly merge, a new management value system is emerging that blends the best of both approaches. In both old-economy and new-economy companies, executives are usually consumed by their commitment to making the company as successful as possible, whatever it takes. (The major difference, as one computer consultant puts it, is that, in the new economy, "You can choose whichever sixteen hours a day you wish to work.") Rapid response time, customer-centricity, flexibility, adaptiveness, tight controls on costs and capital investments, global reach, outsourcing—all of these have become standard business practices for companies in every industry, reflecting the old/new blend that increasingly dominates world business.

One final rule that has carried over from the old economy into the new, and that is still as valid as ever, is this one: *Make yourself a profit center.* This means, on one level, that you should strive to *grow* rather than *drain* your company's assets—to somehow attract or produce more revenue than you expend. Beyond that, it means that you need to understand the profit model that drives your company's business—how money is made, and how your job contributes to that.

If you personally are a source of profits for your company, you will be virtually indispensable. Aspire to have people say about you, "We can't fire *her;* she makes us too much money!"

TODD'S STORY

The tech sector generally likes and rewards bold behavior. And the fast-changing environment of the high-tech world means that sometimes you have to act first and think about it later. By contrast, Todd thought first, acted later. Here's his story.

> **Todd is a vice president at a company that provides seed money for start-up firms. He heard that the chairman of a multi-billion-dollar company ("Dynamox, Inc.") was going to be speaking at an industry meeting. This offered an unusual opportunity: Despite his company's fame, the chairman was not well known, because he kept a very low profile, letting the company president act the role of the media darling. Todd decided to use this event as a way**

of meeting the chairman. Because Dynamox and Todd's company operated in related fields, such a connection could be very valuable.

Todd scoured Dynamox's Web site to find out what he could about the chairman's background. He read the chairman's speeches, downloaded his photo, and even learned that he didn't drink alcohol.

Todd arrived at the meeting early and watched the crowds gathering. When the chairman walked in, he was alone, and no one talked to him, not even the executives from his own company who were present. The chairman walked over to the bar, got a bottle of water, and stood quietly apart from the others.

Todd walked over, introduced himself, and mentioned a couple of things he remembered from the chairman's speeches. They talked one-on-one for the next thirty minutes.

Later, one of Dynamox's vice presidents, a friend of Todd's, asked him, "Who was that you were talking to?"

Todd was surprised. "That was the chairman of your company."

Todd and the VP went on to discuss a possible deal between their two firms. After a few weeks, when discussions about the deal got to a point where Todd knew the details would be heading up the chain of command at Dynamox, he e-mailed the chairman through his assistant. In his message, he reminded the chairman how they had met, told him that he had previewed the deal he'd proposed, explained that a meeting was coming up, and offered to answer any questions the chairman might have.

The next day, the chairman called Todd. The deal—an extremely lucrative one for Todd's firm—was as good as signed.

Here are the steps Todd took to achieve success.

- He was in the loop to learn about the event.
- He knew what he wanted to get out of it.
- He did his homework prior to going.
- He planned his approach.
- He took the initiative.

- He was affable and interesting for the chairman to talk with.
- He listened and learned from the conversation.
- He followed up and ensured the outcome.

Todd courageously took risks, including the following:

- Todd did research. (At best, he risked wasting his time; at worst, he risked appearing nosy.)
- Todd watched for the chairman's arrival at the meeting. (He risked appearing calculating.)
- Todd went up to the chairman and initiated a conversation. (He risked rejection.)
- Todd talked with the chairman for thirty minutes. (He risked being pushed aside for others.)
- Todd followed up with an e-mail. (He risked looking pushy.)

Although there were risks, Todd believed they were worth taking. The same kind of approach, and the same lesson, applies to job-hunting, client development, and networking of any kind. If you hope to be successful in building your own high-tech career, learn from Todd's story and imitate him.

5

MAKE YOURSELF KNOWN

"Good networkers try to make a lasting impact on every person they meet, whether through something they say, something they do, a compliment they give, a bit of help they offer, or just a moment of shared laughter."

—David Klotz, founder of Comocean

Being in on the Buzz

In the tech sector, a lot is happening all the time, all over the world. The newest, latest, hottest is what the buzz is about. So the buzz is what you need to be in on, because when you're in on it, you see where you fit into it.

To be in on the buzz, you have to be out where people are talking. You can't wait to read about the buzz; by the time it's in print, everyone else knows about it, too. You want to know the latest before others do.

This chapter is all about being in on the buzz—about how to make certain that you know what's happening and are known by those who make things happen. This doesn't mean you should engage in idle gossip or grandiose ego-boosting; instead, it means you should go about the hard, important work of being plugged in to the centers of power and influence in high tech, and especially in the specific business you're part of.

The right people are those who are producing results at the highest levels of business—the ones with the power and money to make a difference. These people are centers of influence. They include executive recruiters and CEOs, as well as the senior executives, venture capitalists, creative thinkers, bankers, entrepreneurs, consultants, and industry leaders that the recruiters and CEOs go to when they want ideas and recommendations. Those are the people you need to know and be known by—the A-list of the high-tech industries.

You don't have to become best friends with the A-list people, but you should at least work to make them aware of your name and your reputation. When someone calls an A-lister and asks, "Do you know anyone who has done X?" you want the response to be "Oh, sure," accompanied by *your* name.

The Power of the Network

You probably already know about the importance of being in on the buzz. It's often referred to as *networking,* and most people are familiar with the term. But many people who know about networking neglect it. Time constraints are a common reason. When you're busy (make that *swamped!*), overwhelmed by deadlines, budgets, and stress, you may not take the time to make and nurture business connections.

Some people feel as though it's vaguely improper and probably unnecessary to make a special effort to know and be known. "After all," they say, "if I do a great job, the word will spread fast enough." It's true, of course, that you have to produce superb results on the job. That's a given. But being good at what you do is not enough. You also need to have the right people know about it—and about you.

A familiar disparaging, cynical comment people sometimes make about someone who is successful is: "It's not *what* you know, it's *who* you know." The implication is that networking enables people without talent to rise to the top. The truth is that no amount of personal charm or skill at making connections can ever make up for a lack of knowledge, dedication, or effort. Networking is an important adjunct to productivity—not a substitute for it. Networking is also a natural and unavoidable aspect of human relationships. Part of what you know is who you know. Just as you need to know the technology in your business world, you also need to know the people in that world.

It's also inevitable that personal connections influence how business decisions get made. Think about it. Suppose a stockbroker calls you out of the blue and asks you to invest some money with her. What is your likely response? Probably a curt "I'm not interested." But suppose a friend from the gym, the church, or your book club calls and says, "I've been working with a broker. He's a straight shooter, and he's earned good money for me. I think you should talk to him." You'll likely be much more receptive to what the stockbroker has to say. The difference is the power of the network.

A successful CEO, John Robert, told me about an earlier venture of his, which floundered. What does John know now that he wishes he'd known before? "I wish I'd known the saying, 'It's not what you know, it's who you know.' We were in the forefront of the technology [we were creating], but we didn't know it. We were too busy doing rather than developing the contacts we needed to help us see and understand the opportunity."

Every company executive, recruiter, and venture capitalist that I talked to in researching this book emphasized the same point. No wonder *The Wall Street Journal* reports that 94 percent of successful job searches happen through networking. As John Challenger, CEO of Challenger, Gray & Christmas says, "Time and again people work very hard and are good at what they do, but they don't know anybody. They put their heads down and spend no time developing social capital. . . . [As a result,] they have a hard time finding a job."

Networking with people in and around your industry is the best way to:

- Gain perspective on yourself and your career.
- Learn about new or unpublished job opportunities.

- Keep up to date with industry trends.
- Meet and get to know your industry's key players.
- Learn about the problems in your industry, and the solutions people are developing.
- Find out about the competition for the best job openings.
- Learn about the new training you may need.
- Discover ways to help others (who may someday be able to help you).

Steven Aldrich, president of QuickenInsurance, sums it up well: "Networking is key. Building a group and staying in touch with the group through all the changes that occur in everyone's lives can be difficult, but it's essential."

How to Get Started in Networking

The first step you need to take in networking may be an attitude change. Many people hesitate to network because they fear they'll be perceived as pushy, needy, or intrusive. The underlying assumption is that you have little or nothing to contribute to the people or the group you hope to network with. That's probably false. In reality, one of the biggest perks of networking is the pleasant discovery that you have as much to offer as most other people in your business. As a result, most people find that, in addition to its other benefits, networking proves to be a huge confidence-builder.

The realization that you have something to contribute to the conversation can help you develop a more positive, friendly approach to networking. Rather than focusing solely on meeting people who can help you in your career, you should think in terms of *sharing*—of a free and equal give-and-take between people with mutual interests and complementary needs. The truth is, that's how networking operates.

How do you get started? The first principle is simple: *Network wherever you are.* Literally any place where there are people can be fertile ground for valuable business connections.

Here's a small sample of places where people I've interviewed for this book have made important contacts.

- The gym, the grocery store, church
- Sporting events, concerts, movies
- Kids' soccer games, school events, family gatherings
- Computer user groups
- Classes and seminars
- Pink slip parties
- College alumni meetings
- Employers' and entrepreneurs' roundtables
- Business exchanges
- Trade organizations and professional meetings
- Chamber of Commerce meetings
- Meetings of Young Entrepreneurs Organizations, Young Executives
- All the C-level gatherings: organizations specifically for CEOs, CFOs, CIOs, CTOs, COOs, CMOs, and those who work for them

There are city, state, regional, and national organizations for almost every possible professional or business interest. Don't pick just one—try a wide range of groups that attract a variety of ages, seniority levels, professional interests, geographic and personal backgrounds, and so on.

Business publications such as *BusinessWeek, Fortune, Fast Company, Forbes,* and *Inc.* sponsor seminars. So do investment banks, business schools, and professional organizations. These enable you to meet people from around the country and the world, measurably broadening your circle of business contacts.

You can even start your own group. Michael Doyle, senior brand director at Immedient, put together a mix of entrepreneurs and high-tech executives who come together to chat, swap stories, and form strategic partnerships. Doyle explains, "Every couple of months we get together after work or for breakfast with a structured agenda." All the members of the group benefit from the connections, especially Doyle.

"Be an assembler of interesting people. Put together sort of a mosaic," advises Jack Shields, general partner with Best Capital Ventures. And Kent Lassman, director of technology and communication policy with Citizens for a Sound Economy, recommends seeking out those with viewpoints other than your own. "Spend time talking to people who don't agree with you," he says. "If they agree, just say thank you, and move on."

Some businesspeople go out of their way to conduct business specifically for the contacts it brings. Kevin King is CTO and president of Precision Solutions, a company with a worldwide customer base. King moved to Longmont, Colorado, because he loved the lifestyle there. But he soon found that he missed the connection to clients, because Precision Solution had no customers there. King's response? "I took a contract locally just to get known, instead of only doing the international work."

You can also make useful contacts through personal, nonbusiness interests. It could be the local homeless shelter, humane society, an organization that helps battered women, or your town's art museum. Find projects you can work on to get to know other people better. "Volunteer for the worker spots. You will not be beaten out by the competition," says Jack Falvey, founder of MakingTheNumbers.com.

"Get out two nights a week," says one recruiter. "Do this for three years and you'll have a powerful machine behind you."

Attending a regular schedule of varied business and social events is essential for any ambitious businessperson. Recently, one of the country's top recruiters showed me his personal schedule for one week. On four days, he had breakfast, lunch, *and* dinner meetings scheduled with current, former, and potential clients. On the weekend, he had two client Christmas parties and a football game (to be viewed from a client's private box). He also had a board of directors meeting, two advisory board meetings, and a speaking engagement. All that on *top* of running his business!

Your schedule may never be quite that crowded. But consider this: If an enormously successful businessman with a wealth of personal connections considers it worthwhile to invest this much time in nurturing those connections, shouldn't you strive to make a comparable level of investment?

Working an Event

There are a number of specific techniques you can use to make the most of any networking event. The best method for working an event starts with careful planning.

Prior to the event, investigate the meeting format, so you know what to expect. Ask about the dress code, so you can dress in a style that won't make you feel awkward or out of place.

Then find out who will likely be there that you would like to meet. Gather some basic information about that person. Look up his or her company or individual Web site to read a biography, personal information, and perhaps something he or she has written. Look for a photo so you can spot your target easily from across a room.

On your way to the event, prepare to enjoy it through positive self-talk. Send yourself messages, such as, "I'm going to make this time really productive and interesting," so that you'll walk into the room feeling confident, comfortable, and competent.

Then, when you arrive at the location, pause, size up the situation, and head first toward the person or group you find most intimidating. Everyone else will be easy to speak to after that. If there is a bar or a buffet, don't head that way first. You may want to go there later as an excuse to get out of a conversation.

During the course of the afternoon or evening, seek out the A-listers attending the event and make it your goal to meet them. For example, most organized meetings have a guest speaker. Make sure you go up and talk to that person if possible. What should you talk about? You can't go wrong by complimenting the person on his or her presentation, then asking a relevant question. Don't be a pest, but don't give up too easily, either. If the conversation strikes any sparks of mutual interest, offer your business card. You'll probably get the other person's card in return.

BREAKING THE ICE

If you don't have an obvious entrée to a conversation with someone, try the following technique for breaking the ice: Walk up to a small group and stand nearby with a relaxed expression on your face. Listen to what people are talking about, and chime in appropriately. You might say, "I overheard you talking about X. Do you mind if I listen?" Then introduce yourself, giving a brief explanation of why you are interested. (For example, "My company is thinking of launching an initiative in the same area.") You may want to interject a comment like, "Good point," or "You're so right," or "Why do you say that?" and volunteer something that could be of interest to the group. Ask some questions, but share relevant information also, so you're not just taking, but giving, too.

Don't hesitate to ask for specific information and introductions. Doing so expresses respect for the knowledge, connections, and experience of the person with whom you're speaking. And offer to return the effort. Networking should always feel like a *mutually beneficial* experience.

Once you've learned what you feel you can from a particular conversation, excuse yourself, perhaps using a line such as, "I think I'll go find something to munch on." Tell them it was nice talking to them and leave. On your way to the bar or buffet, look for the next person or group you want to talk to.

Christina West, founder and COO of Cater City, has her own ice-breaking technique. "I go [to events] with a business associate from a different company. Then I can introduce him to people and say good things about him he'd never say about himself. And in turn he can introduce me and say good things about me that I wouldn't feel comfortable saying. We provide each other instant credibility. We meet twice as many people, plus we can rescue each other from conversations we want to end." Simple? Yes! Effective? Yes!

After gathering five solid new pieces of information, you can leave the event if you must. When you get home, write notes on the back of business cards, jot down anything you committed to do—and do it.

Recently I was at an international conference at the convention center in New Orleans, which seats more than 10,000 people. The place was full of people eager to hear a particular CEO. He spoke eloquently, humorously, and impressively, and was received with huge waves of well-deserved applause. Then he walked off stage, behind the curtains, as the enormous audience gradually filed out of the auditorium.

Now this was someone I wanted to meet. So I walked up the steps onto the stage, ducked behind the curtain, and found the CEO there, standing by himself, drinking a glass of water. I walked up, complimented him on his speech, introduced myself, and asked to exchange cards, saying, "I'd like to contact you on a project I'm working on that may be of interest to you." He thanked me, we traded cards, and shook hands. About this time, one of the CEO's staff "handlers" arrived and whisked him away.

There were 10,000 others who had the opportunity to meet Mr. CEO just as I did—but they didn't. Most people are afraid to talk to strangers, particularly "important" strangers. Instead of fearing rejection, you should take advantage of the odds and seize the chance to meet people you'd like to know.

Harvey Mackay, CEO and author of *How to Swim with the Sharks Without Being Eaten Alive* as well as several other books, told me a similar story about how he met ABC News anchor Ted Koppel. Koppel was scheduled to speak at a Chamber of Commerce event that Mackay was also attending, and Mackay simply volunteered to pick him up at the airport and bring him to the event. Typically a car service would have been used, but why let the driver get the pleasure of meeting Koppel?

Mackay didn't leave it there. His research had showed that Koppel was a big tennis buff, so Mackay just happened to leave his own racket lying on the backseat of the car. Of course, Koppel inquired, "You play tennis?" Mackay had been an amateur championship player and belonged to the best tennis club in Minneapolis. The upshot? Mackay and Koppel found time between getting from the airport to the speech to play a quick set at the club!

Making One-on-One Connections

Not all networking revolves around public events. Jim, a very successful senior executive I talked with, said he had gone to networking events no more than three times in his fifteen-year career. Yet when I asked him how he'd approach getting a new job, he said, "I'd make calls to my contacts."

"So where do your contacts come from?" I asked.

"From connections through work," Jim replied. "I've developed relationships with lots of people in companies where I've worked." He went on to explain that he does some sort of relationship-building every single day. "It's my self-assigned job. My network of sources includes current bosses and colleagues up and down the ladder, and former bosses and colleagues inside and outside of the organization. And the business is broad enough to provide fresh contacts globally. One company I worked for had 450,000 employees with 160 separate operating units. Another had 200,000 employees. When people I've worked with leave to join other companies, I stay in touch. That broadens my base outside my company." It's a great idea, especially in a high-turnover world like that of today's high-tech industries.

If you work for a large company, as Jim did, you have a lot of options internally. But even with a small company you have important customers, vendors, peers, and investors—all of whom can become sources for you.

One way of meeting a businessperson you're eager to know is simply to call the person up directly or write a letter to introduce yourself. You could literally select someone from the front page of tomorrow's *Wall Street Journal* to contact. What's the worst thing that could happen? He or she may not respond. So what? You won't be worse off than you are now.

Obviously, it'll be all the better if your phone call or letter offers something to cause the person to want to respond—a reference to a shared interest, for example, or a business area in which you may be able to contribute knowledge of skills he or she would find useful. If you do, you may receive a response, whereupon you can follow up appropriately.

In an age where technology makes it easier to e-mail, page, or reach someone by cell phone, you should make the extra effort to meet the person face-to-face. Someone who has never met you in person would be far less willing to refer you to someone else.

Seize every opportunity to get "face time" with your networking targets. Keith Johnson, a vice president at ThinkArena, lives by this rule. "If I'm invited to a dinner where important people will be," Johnson says, "I do not turn it down. And it's not just formal events. . . . I discovered that Turley's is the place in my city where executives go to breakfast. I got photos of company executives I wanted to meet and studied them so I'd recognize them. Then when I saw them eating in there, I could go up and introduce myself."

If one of your targets invites you visit him or her, you'll want to handle the meeting so as to maximize its value. Here's how.

- In advance, to make the meeting mutually productive, list the key topics you want to the discuss and the questions you want to raise. And do your homework by reading all you can about your host's background, interests, and views. This will make it unlikely that you'll waste time (and annoy your host) by asking questions whose answers you should have known.
- Arrive on time. If you're a little nervous, take a deep breath to relax before you enter. Pause in the doorway a split second just to breathe deeply one more time and slow down. Don't race in. Walk in with good posture, eyes on the person, and a relaxed smile on your face. Shake hands. Pause, don't rush to sit down. See where he or she goes first, then position yourself accordingly.
- Say, "It's great to get to spend a little time with you." If the person doesn't know much about you, give him or her a three-sentence summary—just the basic facts about your background, experience, and interests. And remind your host about what you want to discuss.
- Don't look or sound hesitant. Engage in conversation rather than interrogation. Chime in with a compliment or questions, or to volunteer some information. If you find yourself rambling, stop talking.
- Listen with an occasional slow, deep, purposeful nod, not a bunch of bouncy bobs of the head. Ask probing questions, such as, "That's inter-

esting, can you tell me more about that? Can you give an example? Where else have you seen that happening? or How do you view . . . ?"

- Contribute things that you know or have learned that might be of interest and value to your host. Not gossip, of course. Your goal is to come across as a peer, colleague, or friend exchanging information.
- Use mild flattery. It's okay to flatter your host on his or her opinions or experience—if you're sincere.
- Even if you already know something your host is telling you, it doesn't hurt to sit and listen without interrupting with a remark such as, "Yes, I know!"
- Finally, stay within the agreed-upon time limit, unless your host practically chains you to the chair. When it's time to leave, thank him for his time and thoughts. Offer to be a contact to call upon if he or she is studying or pursuing some area with which you can help. Then leave.
- After the meeting, send a follow-up note, voicemail, or e-mail. In a couple of weeks, touch base again to offer an update on anything you're doing or learning that would be helpful to your new contact.

Nurturing Your People Skills

One of the benefits of networking is that it's a great way of improving your interviewing skills. The logic is simple. If you want to become a better actor, *act*. To become a better writer, *write*. And to become better at interviewing, *interview*. Networking conversations are like low-stress, high-impact, self-initiated interviews. By having lots of these mini-conversations when you aren't under pressure, you get better at explaining what you want people to know about you. And the more you do it, the more skilled and focused you become.

"But I hate small talk," people sometimes protest. Then don't engage in small talk. Talk about things of interest to you and to others. One surefire way of feeding a conversation is to try to discover what's of keen interest to the other person, then talk more about that. Offer some helpful ideas. Don't assume, "Oh, she's probably already thought of that." Maybe not. And of course you can steer the course of the conversation by inserting information about your own interests also.

Pay attention to what people say. Care about what they say. Listen hard, and practice reading between the lines. As management guru Peter Drucker says, "The most important thing in communication is to hear what isn't said."

REACH OUT AND TOUCH SOMEONE

Physical touch is a communication tool few people take advantage of. It can help you bond with people, not just communicate. Yes, you can touch strangers at a networking event if you do it with the right attitude and technique. Here are some tips on effective use of touch in a business context.

- Reach out and touch the person you're talking to. Maintain physical contact for a split second as you speak directly to the person.
- Place your hand and remove it in an equally purposeful and definite manner on an acceptable part of the body: shoulder, elbow, forearm, hand.
- Relax, smile, and look as if you expect the other person to accept the touch in the manner you gave it (supportive, encouraging, caring, respectful).
- Do not use touch with any intimate overtones. Be sensitive to the person's reactions.
- Above all, be consistent. Use touch with men and women, young and old, the likable and the not-so-likable.

Even if you're skeptical about this, try it to test the effect. You may be amazed. Five minutes effective communication accompanied by the appropriate touch can be worth five months of hard work.

A LITTLE GIVE AND TAKE

Another secret of effective networking is to give at least as much as you take. If you only take, you'll get a reputation for that, and in time people will avoid you. If you give—especially if you give first, without knowing whether or not you'll receive—people will be attracted to you. "An offer of reciprocity gets my attention," says Alan Grafman, CEO of Modelwire. "It's my personal secret to always ask the person I'm networking with, 'How can I help you?'"

Why not try this technique right now? Think of a couple of people that you know who might benefit from knowing each other. Call them up, and explain that you think so-and-so would be beneficial for him or her to meet.

I recently wanted to use the professional services of an artist I know. I didn't feel I could afford his top-drawer price, but I wanted his top-drawer work. So I explained that to him. I added, "I know you're worth it, I just can't afford it at this time." Then I volunteered, "I know someone who could use your services. When we finish talking, I'll call him and suggest that you two meet." And I did. The second person then called and made an appointment with the artist to discuss some business. Later, the artist called me and said, "Thanks for the introduction. We're meeting next week. And don't worry about the price for your project, I'm going to give you what you need for a price you can afford."

The more you do to help someone else's career, the more willing that person will be to help yours. You know the expression, "What goes around comes around." In the high-tech world, what goes around comes around even faster.

By the way, if you're a "gray"—older than forty—make an extra effort to get to know and network with young people, who tend to be more in tune with new trends in technology. In return you can make the relationship mutually beneficial by providing insights and advice based on your years of experience in the tech world.

What if you find your efforts to network stymied by intense shyness or anxiety? Most of the time, you can reshape your behavior and thought patterns to control and overcome shyness. Everyone has some degree of social phobia; most people feel nervous meeting and talking to strangers. If you are excessively shy, you have to deal with it. The truest and best way is to understand that others experience it too. Other people feel just as nervous as you do at times, maybe more so. So if you bravely act first and help those around you relax, you'll get more out of your time together. This doesn't mean you need to become a social butterfly to network successfully; it does mean that you can't be lazy about making strong, diverse connections on an ongoing basis.

Sometimes it's necessary to draw limits on your networking activities. As you get better known, you will be the person with whom others will want to network. You may have to guard your time. A simple statement such as, "If you have any other questions, just send a note," can be a helpful way of ending a conversation and giving yourself a chance to move on. If you're pressed for face-to-face time, you can say, "I'll be happy to spend a few minutes on the phone with you."

Larry Feld, president of Human Performance Strategies, has developed his own techniques for controlling his networking time. "I qualify [people] up front and then determine how to best manage the situation," he explains. "I inquire about the subject, assess how much time is required to deal with it, whether on the spot or whether to schedule a more convenient time. If someone wants me as a source but I don't want to be, I'll say, 'I am not comfortable in making such a recommendation because I'm not that familiar with your work.' "

Keith Johnson, a vice president with ThinkArena, adds this advice: "If people are blatantly trying to use me to get themselves somewhere, sometimes I'll still direct them to the right people. But I won't give the kind of introduction that they would get if they had at least faked some friendship or mutual respect. . . . And when people send others to me, I'll almost always talk to them. You just never know."

Staying in Touch

Wherever your connections come from, it's important to periodically touch base with them. It takes less than five minutes to reach out in the following ways.

- Phone and comment on research, an article, a book, or whatever that you found that would be of interest.
- Send a copy of research, an article, or a book.
- Phone to give a compliment on a business success.
- Send an article that you wrote.
- E-mail an inquiry as to whether the person was affected by the flood or snowstorm that hit his or her area.
- Contact the person's secretary or administrative assistant with something of benefit.
- Phone or e-mail to ask for an opinion on something you're working on.

You get the idea. The list is endless.

About once a month, I receive an article of interest from Paul Schlossberg. He's the CEO of D/FW Consulting, and he travels a lot. On his flights, he seemingly reads every magazine published, and he constantly clips articles to send to people in his community of contacts. One of his secrets is that he has envelopes pre-addressed and stamped in his briefcase, so when he sees an article, it takes about three seconds to send it out.

Another person who knows how to make himself a valuable source is Eric Weissmann. When I finished interviewing him for this book he asked this simple but great question: "Is there anyone you still need to interview for the book? Is there someone I could introduce you to?" Comments about such assistance is a sure way to win someone over.

Sometimes networking relationships become a little closer than usual. Lisa Mandell, vice president of marketing at siteRock, says, "At each job, I've found a few people internally who I could trust. I call them my confidants. They were people I could say anything to. I talked to them a lot. Today I try to be that person for others."

Mandell says that *every* time someone asks for her help or advice, she goes out of her way to give it to them. Every time a headhunter calls, she "always, always" returns the call and gives a referral. "I'll even refer people

I don't know if I think it will help them." She adds, "I always ask how the recruiter found me, too." That way she can follow up with the person who passed along her name.

People like Schlossberg, Weissmann, and Mandell really get it.

As you can see, although you may not be able to go to networking events every week, there are plenty of ways to create connections on your own daily and weekly.

Long-Distance Networking

There's no substitute for face-to-face conversation, but your networking efforts *can* be done long distance—it just takes a little more effort and planning.

One technique is to pre-organize face time during a concentrated period of travel. For example, suppose you live in Cleveland and want to work in California's Silicon Valley. Do all your research in advance, make contacts, and fly out for a week in California. Set a goal of three meetings every morning and three meetings every afternoon, and in the evenings attend the San Jose or California chapter of groups you belong to or might want to belong to. (This kind of intense schedule won't be easy to maintain. With the traffic congestion typical of Northern California, you'll have to do some fancy coordination, but it is possible if you think it through and plan your meetings with a detailed local map in hand.) Add a job fair to your agenda while you're there, and that will give you a lot of face time.

For some people, long-distance networking is a regular necessity. I used to be one of those people. A number of years back, I fell in love with a mountain man—a real one. He managed a ranch sixty miles from town up in the Colorado mountains. When we married, that's where we lived for three years. Being self-employed, I could work from anywhere via phone, fax, e-mail, and snail mail, but the geographic location was not conducive to a lot of face-to-face meetings. I frequently traveled to give speeches and do consulting, but otherwise I was geographically out of touch.

I had to do what I could to maintain contacts through other means. That meant planning phone calls; figuring out justifiable reasons to touch bases; maintaining more regular and attentive mail contact; and seeking more creative, memorable, and productive contacts in *everything* I did. The interesting outcome: In those three years, despite my geographic constraints, thanks to a concerted effort to provide good work to clients and really keep connected, my business volume doubled every year.

WORKING THE PHONES

Here are some tips for networking by phone that surprisingly few people follow.

- If it's an appointment phone call, be on time. When you get the person on the line, confirm that it's still a good time to talk.
- Have your questions spread out on the desk or table in front of you.
- Have paper, a pen, and a spare right there. Don't type notes onto your computer—the other person will hear the dull tapping, and that can be a put-off for some people.
- Instead of using a speakerphone or clasping the handset uncomfortably between your neck and shoulder, use a headset.
- Don't speak too fast or too quietly. Smile when you talk. Even though the person can't see, your voice will sound more friendly.
- Make phone calls standing up whenever possible. Your voice will sound stronger and clearer.
- Start by stating why you're calling. Remind the person who you are and how you met, and mention what you want to accomplish in the brief phone time together. Ask if he or she has a few minutes to talk. Be clear and concise.
- Stick to the agreed-upon time frame.
- End the call by asking if you can call back or e-mail with another question if one comes to mind. Then do so. It shows that you found the person helpful, and the recognition will be appreciated.

Technology—phone, e-mail, fax—provides wonderful tools that can help people stay in touch. But remember that no form of technology substitutes for pure human contact. As Claude Pineault, CEO of the Quebec Software Promotion Centre, puts it, "You get your contacts from your personal web, not the virtual Web."

MARION'S STORY

What follows is an account by Marion Martinelli, in her own words, concerning her use of networking as a powerful means of expanding and advancing her career.

> **I am always looking to meet new people, to learn about their industries and the occupations they hold, who they are connected to, and how I can, through these relationships, jump on opportunities that might not otherwise be available to me.**
>
> **I always keep an updated resume and reference sheet in the trunk of my car. You never know who you might meet. For example, my last employment opportunity came from a man that I met in a cafe. He shared with me the fact that he'd just been hired at a progressive new start-up company in Palo Alto, California. I took advantage of that opportunity by handing him my resume, and he offered to pass it along.**
>
> **Approximately six months later, there was a message on my voicemail from him. He said that the company had grown quite a bit and they were looking for someone with my skills. I got the job.**
>
> **The man who helped me to get into the company lives in my town, a good two hours away from our place of employment. I would have had a heck of a time getting into the company if I hadn't known someone. In fact, everyone who works at the company was recruited because they knew someone who already worked there. No joking.**
>
> **Good people attract good people, and I think that may be why some of the fast-paced, start-up companies try to hire the best—and then hire all of their friends and family.**

That man is now at another start-up. We still keep in touch. Through that relationship, I met another manager at another start-up company. It's like a chain reaction.

Never underestimate what can happen to you almost anywhere. Cafes are full of potential. A couple of months ago, I was sitting in a cafe and talking to a friend about my employment challenges. The man sitting next to us overheard our conversation and handed me his business card. He was a project manager for Hewlett Packard. I sent him my resume and he passed it along to some of his colleagues.

Plant those seeds. You never know when they will grow. I have potential opportunities planted all over the place. Just recently, I was at the local Starbucks, reading the San Jose *Mercury News,* and I was so enraged by one of the front-page stories that I turned to the man sitting next to me and said, "I just have to share this with you!" We got to chatting, and it turned out that he was a manager at a start-up company that is ready to go public in a couple of months, close to where I live. He took my resume and passed it on to the VP.

It's so important to latch on to and develop relationships with people who share your values, goals, and professional ethics. When you are not part of the pack, it is extra important to keep in touch with the few who will support you professionally.

The term "networking" has really been reinvented. In the old days, networking was a word that the majority used to recruit and promote white males. It was thought of as a secretive political technique used solely to keep out women and minorities.

Networking no longer carries that negative meaning. Over the years, along with the changes in business and the workforce, networking has evolved and it is now viewed as a window of opportunity, a connection to the next big career move.

Today, networking is a necessity, a communication link, used for career development, sharing knowledge and resources, keeping abreast of current market trends, and creating alliances. With the rapid changes in technology and the necessity for more efficient channels of communication, I feel that networking will evolve into a science.

6

LAUNCHING YOUR JOB SEARCH

"You not only have to be good at what you do, you have to get to know the right people and maintain relationships with them. Those people want to get ahead in life also. If you can help them, they will open doors for you."

—BILL MANNING, senior electrical engineer, vertical motion simulator, NASA

Your Network of Sources

Sources are people who act as referrals, references, idea generators, idea testers, insight providers, mentors, advisers, intermediaries, supporters, and confidants. In short, they are your friends in the sometimes unfriendly process of looking for your next great job.

Your network of sources is what separates you from the pack of other people who want to get into high-tech industry or move up the ranks. In one form or another, we're all familiar with the power of the source. Here's a simple example. During an interview on *60 Minutes II* near the end of his presidency, Bill Clinton casually recommended David Herbert Donald's biography of Lincoln. That day, the book—already a strong seller—was ranked 6,049 on Amazon.com's sales index. Two hours after the interview aired, it had skyrocketed to 297. *That's* the power of the right source.

With a great source, you soar past the competition and capture the attention of someone with a compatible need or interest—just as Donald's *Lincoln* broke out of the pack of thousands of other historical books after Clinton praised it.

What transforms an acquaintance into a source? The number-one reason someone significant will be willing to help you in your career is because someone he or she knows and respects suggests it. Other reasons may include:

- To return a favor
- To appear important and influential
- To enhance personal power
- To create or repay a personal debt
- As a gesture of pure friendship or generosity
- Being pestered into it

Of course, the last reason is the least desirable. Don't go to the lengths of becoming a pest in order to gain someone's help. But notice that all of the motives, in one way or another, stem out of self-interest. People who serve as sources for others are doing it, in the end, because they *want to* and because it *benefits* them, directly or indirectly. Therefore, you shouldn't feel uncomfortable about enlisting others' help in your career efforts. They'll assist you to the exact degree they want to and that they expect it to benefit them—and no more. (And, of course, through the professionalism and courtesy you show in the process, you'll be communicating to your source the fact that you expect to serve them in the future, as one of their sources or as a source for someone they care about.) Thus, the concerns many people feel about "using" others in their job search process are overblown.

Think about how it works. Suppose you know Alexa, who is currently president of her own company. You've worked with her and consider her a friend. Then, in conversation with another friend—Greg—Alexa's name happens to come up. It's only natural for you to comment, "I know Alexa—we worked on the XYZ project together, and she's great."

The following week, you run into Greg at an industry event. Greg comments, "You said that you know Alexa. I'm going to be sending her a letter about an opening my company has for an executive vice president. Would you mind if I mention your name?" You respond, "Of course not. In fact, when I talk to her next week, I'll tell her to expect a letter from you."

You haven't been "used." No one has put the strong arm on you. In fact, it feels pretty good to help two people you like get together. If their relationship develops well, it can only benefit both of them—and you.

Lisa Mandell, vice president of marketing for siteROCK, says it well: "I never feel used when people ask me to be a source for them. People should never worry about that. If someone doesn't want to help, why would you want that person anyway? Most people remember when they needed help themselves, and they are happy to give it."

Of course, there are limits to the kinds of help you'll ask for, limits set by your own professionalism and tact. "Everybody likes to help. It's flattering to be asked," says Tom Keller, CEO of Net Infrastructure. "But that doesn't mean helping with something that the person could have done or should be able to do on his or her own. I have no patience for that. But if it's something that this person has already given 100 percent to, it's really great to help the person extend himself or herself."

Source developing is, in a sense, simply a process of surrounding yourself with good people. When you run a company, this is a must. But why shouldn't the same principle apply to your career? In today's constantly changing business world, the only constant may be the relationships you develop. It's the part of your career mission where arguably the greatest effort should be made, but where often the least effort is made, until it's too late.

Transforming Your Network into Job Sources

The first step in launching your high-tech job search is to begin the process of transforming the network of contacts you've been developing into job sources. What do you need to do to transform your network of friends and acquaintances into job sources? Here are some basic principles to follow.

- **Work at developing relationships when you don't need them.** If you do, you'll be able to comfortably call on them when you do need them. Once you've developed contacts, be sure to do something to cross their mind at least once a quarter. Sources remember you when you give them reason to. This creates a reputation that speaks about you even when you aren't there.

- **Think in terms of reciprocity.** If you live long enough, you'll likely be able to help everyone who has helped you. Someday you may even be able to help their children, which they'll appreciate even more. Ask or observe what would be of interest to your source, and provide it. Others will be much more willing to help you (repeatedly) if you do for them even when they don't ask.

- **Get to know the support people—and treat them with courtesy and respect.** After your source's administrative assistant Alice has helped you, ask Alice for her last name. Why? Because Alice and others like her are the keys to the fortress. If you have the full name of junior support people, you can reach a live person to attempt to figure out what is going on and where the key people you are trying to contact may be hidden.

- **Get to know the gurus in your current company.** One sure way to advance your career is to impress the best people in your company with your knowledge, people skills, and willingness to learn. The gurus are likely to get the high-paying jobs within your own company or elsewhere. If they move on to a competing company, there is a good chance that they'll need more help—and if they do, you want them to think of you first.

Dale Douglas, a successful electrical engineer, has used this technique and explains how it works. "How do you impress the [guru] with your technical skills? It's easy. After you introduce yourself, you get to know the guru. Offer to work on or help them with anything that comes up. When he or she asks for assistance, you oblige. If the guru asks for one page of information on Friday, you provide three on Wednesday. If you build some amazing software, be sure to offer, or at least show it, to the guru. When the guru gets hired at a new company, be assured that he or she will remember you and your enthusiasm."

- **Nurture your sources at other companies.** As we've discussed, referrals from current employees are a major source of many companies' new hires. Juniper Networks is a typical example. "Everyone who joins the company is expected to bring an additional person," says CEO Scott Kriens. As a result, about 90 percent of Jupiter's new hires come by way of referrals.

The same is true at Microsoft. Markus Schweig, a manager there, comments, "We like to hire great individuals who have family—brothers, sisters—who are also skilled and intelligent and would like to come to work for us too."

Companies in need of top talent have established incentive programs to encourage employees to bring in their family, friends, and acquaintances. For a referral who gets hired, an employee may receive a bounty ranging from $1,000 to $10,000 or a special gift, such as a free ski weekend, a visit to California wine country, airline tickets to anywhere in the United States, or even a chance at a Ferrari. According to *Forbes*, Intuit has paid $5,000 per new hire, while a resume gets you dinner and a movie even if no hire is made; Respond.com gives a five-day luxury trip to Hawaii for an in-demand engineer; and Intel holds monthly drawings for digital cameras and DVD players in exchange for resumes.

- **Treat every connection with respect.** When someone makes an introduction for you, treat both the old and new contacts with respect so as to reflect well on the person who offered your name. If you don't, that individual will never refer you again.

I recently told a headhunter friend about a woman I'd met named Dawn. Dawn had great technical skills, a fine education, a well-connected family, good people skills, and a professional presence—the whole package. The recruiter liked what I said about Dawn, talked with her on the phone, and shortly thereafter offered a job. But Dawn never returned any phone calls or e-mails, nor did she respond to the offer letter. Imagine my embarrassment when I heard about it.

It turns out that Dawn had different dreams and was knee-deep in working to develop her own start-up company. She wasn't deliberately being rude but was utterly focused on her goal. Nonetheless, Dawn's behavior was a mistake. By failing to follow the simple courtesy of calling to explain, she probably lost the chance of being referred by me again. What's worse, my recruiter friend will remember what happened, too. If Dawn or her new company ever need a favor from a top-flight headhunter, it's unlikely they'll be able to call on this recruiter.

- **Close the loop.** It's not enough to launch a connection in the right way. To get the full benefit out of a business relationship, it's important to close the loop. Otherwise, the impression you leave may be a less-than-professional one. Here are a couple of stories that illustrate what I mean.

How Gary Did It

I recently received a letter from a senior vice president at a blue-chip consulting firm. Gary did everything right in contacting me. He named a mutual acquaintance who had suggested he contact me. He described the firm he was currently with, and explained that he was moving to Colorado to get married, so his time would be divided between New York and Fort Collins. Gary's letter concluded, "While I'm not sure exactly what would come out of it, it might be interesting to meet for a cup of coffee at some point, since Pete tells me you're someone I should know." There was no pressure for a job referral, just an invitation to chat.

Gary attached a fact sheet on his background—not a resume—including a page of representative clients/assignments and a company brochure. His package was so well done that I did call him, and we planned a meeting.

Ed's Approach

About the same time, a company president I'd once worked with contacted me. Ed's company had merged with another, and he was going to be out of a job. Ed's letter was well done. He wrote, "You may be aware that I have chosen to pursue opportunities outside of those offered to me by XYZ subsequent to their purchase by ABC. I enjoyed working with you during my time at the company, and I am sure that our paths will cross again in the future. Meanwhile, I would appreciate your support in my job search. I am seeking an opportunity as a CEO, COO, or president of a dynamic, consumer-oriented company that will fully leverage my ability to drive positive business performance and cultural change. To the extent that you know of or learn of relevant opportunities or individuals that I should contact or pursue, I hope you will give me a call." To add extra personalization, Ed handwrote a P.S.: "Please keep me in mind as you rub elbows with the 'big guys'!"

When I received Ed's letter, I immediately telephoned him and left a message saying that I thought his written presentation was good, and that I certainly would recommend him should the opportunity come up.

Ed never called back. He had done a great job up to that point—then dropped the ball.

Follow Through

Sources are busy people with many demands on their time. You owe it to them to make it easy for them to help. Ed could have called back and said, "I'm really interested in working for company 'A.' Do you know anyone there?" If he had, I would have responded, "No, I don't know, but I do know the chairman at Company 'X.' Why don't you call him, use my name, and see whether he knows what is going on at Company A? He may have someone you should talk to there."

That last bit of effort is where most people's job search breaks down. To close the loop, you must seize the opening offered.

Making Contact with Companies Directly

Once you've energized your network of contacts, there are many ways of developing leads for jobs. One method, working through executive recruiters, is important and complicated enough to deserve its own chapter. (See Chapter 7.) In the remainder of this chapter, we'll consider some other ways of finding and pursuing specific high-tech career opportunities.

You may well find that directly contacting a targeted company offers the most control and potential for success of any job-search method. It's a method that is largely dependent on timing. It takes a bit of luck—or an exceptionally effective network of contacts—for you to send your resume to a company the very week a great job for which you're qualified happens to open up. I don't think I've ever heard of a situation where someone sent a resume to a company and the recipient said, "Wow! This person is so wonderful, let's fire the employee we currently have in that position and replace him with this person." (And even if this were a realistic possibility, you might not want a job obtained under such circumstances.)

Still, contacting companies directly is the fastest way of getting a Yes or No answer. And when you find the top person in your functional area at the target company and contact them directly, you show an initiative and interest that is flattering to the company. Many great long-term career relationships have been launched in this way.

Some enlightened companies make a point of scanning the incoming resumes with intelligence and care, even when there are no current job openings to fill. Ragesh Jolly, vice president of marketing at Budget Rent-A-Car, takes a personal interest in each day's mail. "If it comes to me with my name on it, I'll read it," he reports. "I browse every piece I receive because I'm always looking for talent, whether I have a job opening or not. In fact, if I find someone who is good but I have no open position, I'll hire him and create a job. . . . If the person seems good but in a different function, like finance, I will send his resume to the CFO in case she's interested."

Today, more and more smart managers are operating according to the same principle. They realize that incoming resumes can replenish their database, which would otherwise rely on recruiting through costly advertisements, executive search firms, or internal referrals. Some conduct continual recruiting through electronic or other means. Ronald Brown, CEO of Maximation, explains, "We have a Web page by which individuals may submit a copy of their resume to us from the Internet. Our recruiters are in immediate correspondence with all individuals, regardless of their experience. . . . Any correspondence sent into our recruiting site not only goes to the recruiters but to me, the president, and several vice presidents. So, in essence, job hunters have contact at the top, even though they don't realize it."

MAXIMIZING DIRECT CONTACT

Here are some procedures to follow to maximize your chances of success when making direct company contacts.

- **Be precise.** The fact is that, as Mark Kimmel, a retired venture capitalist, puts it, "Most blind resumes are received with the same eagerness as people coming to you seeking money." So at least maximize your chances of receiving a friendly reaction by writing to the appropriate person. Human resource people tell me they frequently do not know the specifics of current or impending job openings. So sending your resume and cover letter directly to a function head is critical. For example, if you are in financial management, check out the company's Web site and find the name of the vice president in charge of finance. Get his or her exact title and the correct spelling of his or her name. A company telephone receptionist is sometimes a good source of this information.

- **Treat everyone well.** Anyone whose desk your resume crosses and anyone who intercepts one of your phone calls should be treated as a VIP. After all, he or she is a gatekeeper who can directly affect your chance of making contact with the hiring authority. Any lack of courtesy in dealing with the gatekeepers is likely to quietly torpedo your potential candidacy.

The executive assistant to one company president explained very frankly how she handles correspondence and phone calls from would-be job-seekers. "If the person uses the president's name when writing, I give the resume and letter to him. If the president thinks it would be helpful to someone, he'll send it on. And if the person calls to follow up, I'll tell them who the resume was sent to. But they never follow up and call. I figure they've sent their resume to a zillion people whose names they pulled out of a hat and hope someone in the company will call back. . . . If anyone is nasty to me, I tell the president about it. I spread the word, too, because these job seekers usually just go down the list of executives to contact." So be warned: Treating one assistant rudely can kill your chances of employment by an entire company!

• **Do your homework.** Most letters and resumes sent without the benefit of a specific referral will generate a form response from the human resources department. A typical form letter of this kind reads as follows.

> **Thank you so much for your interest in Semiconductors, Inc., and for sending your resume. It has been placed in our files, and if there is a match between your skills and a job, you will be contacted directly. Your resume will remain active for one year, and as jobs open up in the future your resume will be matched against new openings. You are also invited to submit your resume online for any position that looks like a good fit for your experience. We have many current openings at Semiconductors. Thanks and good luck!**

Of course, it's disappointing to receive a letter like this. But note the sentence, "You are also invited to submit your resume online for any position that looks like a good fit for your experience." Translated from recruiter-speak this means, "If you investigate our job openings yourself, you might find a job in the organization for which your background is perfect. But the reality is that we are too swamped to analyze your resume and compare your qualifications with our needs. So unless something on your resume hits us between the eyes, don't sit by the phone waiting for our call."

The lesson: Launch your own investigation into possible job openings at your target companies. Study the company Web site, which often contains current job listings. Scan the classified newspaper ads, including the so-called blind ads—even though these do not identify companies explicitly, you may be able to guess the names of the firms from the descriptive details provided. Ask members of your network to keep you posted about job openings they hear about. In addition, study the organization charts of companies in which you're interested, using the information you glean to estimate the number and kind of job openings that are likely to be available at any given time.

One human resources director offers this advice: "If I myself were looking for a job, I'd use my network to identify potential companies that I'd like to work with. I'd get the inside scoop from my personal contacts. I'd use the Internet to research companies, learning who the officers are, and if I know someone who might know one of the officers, I'd ask for an introduction. I'd talk to headhunters I know. And then I'd send a resume and follow up."

- **Don't expect a lot from HR.** You might think that, in today's competitive talent market, company recruiters would be super savvy in their approach to attracting employees. Unfortunately, this is not necessarily so. Don't expect company recruiters to do a good job of studying your background and matching it to possible career opportunities. Most human resources departments have way too many other assignments on their plates, including the responsibility for developing and administering programs related to compensation, benefits, employee retention, equal employment opportunity, occupational safety and health, and many other issues. It's no wonder that studies show that the average human resources professional spends 80 percent of his or her time doing paperwork related to current employees rather than finding new ones.

There are some company recruiters out there who are extremely creative. They routinely scan the dot-coms that are in trouble, using them to build a prospect list of job candidates. According to *Red Herring* magazine, "Death Vans" show up in the parking lots of failed Silicon Valley companies with opportunistic recruiters waiting inside to accept resumes. Others have even been known to recruit at rock concerts, baseball games, and bike races.

In general, however, you should expect to do most of the job search and qualifications matching yourself rather than assuming that company recruiters will do it for you. In many cases, the recruiter's role is primarily a negative one. "We are really looking for 'disqualifies,'" explains Doug Heard, a recruiter at Raytheon. These are specific candidate attributes like an unwillingness to relocate or an out-of-line salary demand that can quickly be used to eliminate that candidate. "Staffing is a de-selection process," Doug continues. "If we're too picky or not fast enough, we'll lose the cycle time." Thus, HR often acts mainly as a filter, letting through a fraction of the resumes received for more serious consideration by staff management.

Karen Thomas, director of HR at Raytheon's Strategic Systems Business Unit, gives a frank explanation of what to expect from her department. "Don't expect a company you target to act on your behalf overnight. Sometimes the requirements for the job aren't finalized. They may not be sure how many people they need and are getting agreement on things internally. So if there's a delay in responding to your resume, don't think you aren't worthy or aren't being considered. Don't get nervous and start calling constantly. I personally will try to be honest and not sugarcoat things, but if the response is negative, don't take it personally."

• **Work with the search firm.** Many jobs are filled through outside search firms (covered in the next chapter). When this is the case, it's essential to work through the outside firm. If you send your resume directly to the hiring company instead, it will be forwarded to the search firm—which will likely toss it. After all, if you end up being hired based on a direct company contact, how could the search firm justify its fee?

Try to do enough research to determine whether your target company is using a search firm. Then contact the search firm directly.

• **Follow up.** When you send a company your resume, follow up. Call the executive to whom you wrote and ask one or more of the following questions.

- Does my resume contain anything of interest to your company at this time?

- If there is no current opening at your company, is my background appropriate for possible future positions there?
- Is there someone else at your company that I should be talking to?
- Is there anyone you know outside your company that may have interest in me?
- Could you offer a word of advice concerning my job search?
- Are there any questions I can answer for you about myself or about the other organizations I've been in touch with?

• **Be patient.** Dan Purdy, a manager in demand/capacity planning for Agilent Technologies, speaks for myriad job candidates when he comments, "For me, the toughest part of job hunting is the time delay between the initial job posting, the interview, and the final acceptance paperwork. In each of my past three experiences with job hunting at Fortune 500 companies, the hiring managers stated their need for an individual to fill the position, yet seemingly dragged their feet when considering a qualified pool of candidates."

Dan continues, "From a candidate's perspective, this lack of timely communication on the applicant's status leaves one baffled. . . . So I keep several irons in the fire and follow up with each person after the interview: a thank-you note one week after the interview, a phone call the second week. If there's no response after three weeks, I cross the name off my list. I figure that if they do not recognize and act on talent when they find it, I probably would become frustrated with the people I would be working with anyway."

Dan's attitude is a realistic one. There are limits to what job candidates can do about the delays they encounter. Following up by phone or e-mail is acceptable, but excessive follow-up is a turnoff. Karen Thomas of Raytheon suggests this guideline: "One call to confirm we received your resume is okay. Give us time to see if it's appropriate for what we're looking for. I may be having to convince a manager to talk to you. I may be talking up your skills. We also may be deciding on the specifications for the job."

When you do call, don't appear desperate. Desperate candidates are undesirable; the obvious question they raise is "What's *wrong* with this person that he's so desperate to be hired?" But do try to get something out of the conversation. You did the prospective employer a favor by letting them have a shot at you; you have a right to some reciprocity. At the very least, ask for a bit of information, such as the current status of the job, the likely schedule for interviewing and hiring, or the number of candidates in the running.

Online Job Hunting

Online job hunting has achieved enormous popularity in recent years—especially, as you might expect, in the high-tech industries. It's a fast, easy, and comfortable way to look for job openings, offering candidates less risk, greater anonymity, and less painful rejection than any other method. It's a suitable way for passive job seekers to research company alternatives, career trends, the number of internal and external openings in their field, and how long it takes to fill jobs. However, an online search should *never* be the sole job-hunting method you use. As you'll see, it has too many disadvantages to be a technique you rely upon exclusively. Here are some of the basic facts you need to know about how online job hunting works.

Estimates vary, but there are currently somewhere between 3,000 and 5,000 different job-search Web sites containing millions of job listings. The best known sites are JobsOnline.com, Monster.com, and HotJobs.com, boasting 875,000 postings among them. Specific job listings are current for only two to three months, so it's important to continually revisit sites if you're engaged in a job-search process of any length.

How many jobs are being filled through online searches? According to *Executive Recruiters News,* at Unisys Corp, Microsoft, and Sun Microsystems, electronic recruiting is the second-highest source of new hires, behind only employee referrals. According to *Fortune* magazine, about 10 percent of new hires in 2000 found their jobs online. Other studies, however, peg the figure as low as 4 percent. And about half of all the people who've used online job-search methods say these methods have "seldom or never" led to an interview for a suitable job.

Generally speaking, e-recruiting is for lower-level and middle-management jobs, not senior executive slots. But a few high-six-figure jobs are now being posted online. Monster.com execs like to point out that their database of 10 million resumes includes 100,000 Ph.D.s and 500,000 MBAs. Recruiters say that about 20 percent of management jobs currently open are posted on the Web.

From the hiring company's perspective, online job searches can be fast and cheap. Goldman Sachs reports that a thirty-day run for a classified ad in a major newspaper costs an average of $3,295, while a thirty-day posting on a major job board costs just $167. Furthermore, the reach of the Internet is broader, instantaneous, and worldwide. Thus, the results are often faster as well. A traditional hiring search takes, on average, three months, while an online search usually takes about four weeks. In fact, the initial contact, interview, offer, and acceptance can conceivably happen in a single day.

Online makes things fast for job candidates, too. Rather than making one hundred phone calls, you can dispatch a single e-mail message alerting one hundred companies that you're interested in making a move.

All in all, it's not pure hype when Bud Denker, senior vice president of sales for ComputerJobs.com, claims, "Companies that place classified ads in newspapers to attract employees are going the way of the buggy whip."

TAKING YOUR JOB SEARCH ONLINE

Here are the basic steps to follow in order to take your job search online.

- **Register.** The popular job-search Web sites all provide registration forms to use in listing yourself as a job applicant. You'll be asked to type in basic information about yourself that may help companies zero in on you as a potential employee. Information you may be asked to provide may include your specific job skills, the location where you prefer to work, job titles you've held or would like to hold, your industry background, the names of your target companies, your current salary and future salary requirements, your willingness to relocate or travel, your minority status, your current search status, your dates of availability, your citizenship, and so on. Registration forms on some niche sites can take up to an hour to complete. Registration is usually free.

- **Applying for jobs.** After you've registered with one or more job sites, you can expect to begin receiving notices about relevant jobs, selected on the basis of the company-selected key words that matched words in your registration form. You can apply for jobs that interest you by e-mailing or faxing your resume to the hiring companies.

- **Consider using online job search tools.** The popular job sites do more than simply list job openings. They also provide a host of job-hunting tools, including programs that help you assess your skills, analyze your market value, polish your resume, and conduct research into companies that interest you. Some sites will notify you of available internships, give you personalized career feedback, and provide simple personality testing, job-search tips, salary comparison tables, and reprints of informative articles. Finally, job sites may also connect you to other resources, such as networking and chat sessions, online classrooms, links to business magazines such as *BusinessWeek Online,* and links to sites that sell skill-development courses. Evaluate all of these online tools in terms of their appropriateness and usefulness for you. Some may be quite relevant and helpful; most are probably skippable. You be the judge.

- **Use company Web sites.** In addition to online sites, there are thousands of individual company Web sites you may want to visit. Seventy-six percent of the Fortune 500 firms post jobs online, and some estimates suggest there are more than 40,000 company-specific job boards currently online. Company sites generally provide copious information on the company's history, future plans, financial performance, and key executives—information you can use to impress an interviewer as well as to size up the company while evaluating whether or not you want to work there.

• **Edit your resume to get noticed.** You must edit your resume specifically to draw attention online. The key strategy is to pepper the resume with key words that relate to the specific skills and qualifications of the job(s) you hope to fill. Be specific: Listing particular programming languages, for example, will produce far more hits than simply saying that you have extensive programming experience. (Chapter 8 will provide detailed guidance on crafting an effective resume.)

Here are some other tips for getting your resume noticed on the Internet, adapted from *The Wall Street Journal.*

- If you have your own home page, post a digital version of your resume with examples of past work experience. Include the word *resume* in your Web site address to increase your chance of being noticed by online recruiters.
- If you don't have your own page, check with professional associations that are relevant to your specialty. Many offer free or low-cost Web space and other resources for posting resumes.
- When creating your digital resume, include plenty of links to the Web sites of your present and former employers, colleges, and professional associations and publications.
- If you have a complex job history, consider creating a simpler version of your resume to send to recruiters or potential employers. Offer to provide a longer version on request.
- Don't neglect the niche job boards in your field. Smaller, highly targeted boards can sometimes be more effective than the big brand-name sites.

THE DOWNSIDE OF ONLINE JOB HUNTING

Although online listings can be a helpful tool for tracking job opportunities, the huge numbers of resumes on the Web also make it easy to get lost in the floods of online applicants.

As a result, some recruiters and job seekers have soured on e-searches. "Internet postings are a joke," says one IT director. "We tried an online job posting at our company and received thousands, I mean thousands, of applications. Most of the applicants had little or no experience. To be honest, we had no choice but to delete the e-mails without looking at them."

Online recruiting reaches large numbers of people very quickly and efficiently, but it doesn't allow for the kind of in-person interaction that can help applicants and companies evaluate one another. Realizing this, many companies are reviving old-fashioned, off-line recruitment techniques. For example, HotJobs.com, which gets 90 percent of its revenue from online operations, has revived the tradition of the job fair. CEO Richard Johnson says, "The job fair gives you a sense of a company that you might not get online." If you attend a job fair, you'll be one of about 5,000 people browsing some 100 booths representing companies ranging from legacy firms like American Express and Kraft to pre-IPO start-ups.

PRIVACY ISSUES

Before posting your resume online, read the privacy policies of the online job boards you're considering using. Make certain that you can prevent unwanted eyes from viewing your resume. Some companies employ "Web scavengers" to check for their own employees' resumes online. In response, some job boards let users block specified companies from seeing their postings. If you'd prefer not to have your boss know that you're in the job market, take advantage of these capabilities to preserve the secrecy of your search. However, mistakes can happen. You should assume that your online resume will circulate indefinitely with few or no controls, readable by anyone. At a minimum, you should avoid using a computer owned by your current employer when job hunting online.

In the end, the huge numbers, rapidity of turnover, ease of access, speed, and openness of online job hunting offer both enormous advantages and significant disadvantages for the applicant. By all means, consider using the Internet as one of your job-search tools, but don't rely on it as your sole or even primary technique.

7

WORKING WITH HEADHUNTERS

"I place them as puppies; they grow up to be top dogs, and I place them again."

—A successful headhunter

The World of the Headhunter

The headhunting business—whether it is known by that name or as the *executive search, recruitment, or employment agency business*—is a significant one. Its annual revenues have grown from $3 billion in 1993 to an estimated $8.3 billion in 2000 (according to Hunt-Scanlon Advisors).

A large and ever-changing array of companies are involved in the headhunting business. Kennedyinfo.com's directory of executive recruiters is a helpful source of contact information, firm specialties, staff sizes, fee structure, branch offices, and other basic data. The firm's latest statistics include 5,250 firms worldwide with 13,200 individual recruiters. Clearly, headhunters play an important role in today's career marketplace.

How Headhunters Work

The typical search firm receives hundreds or thousands of resumes every day of the week. (I was with one headhunter when she opened that day's e-mail. There were over 2,000 messages!) Recruiting organizations also receive hundreds of unsolicited phone calls every week. Thus, they are inundated with messages and solicitations from would-be job candidates. Yet with few exceptions, headhunters are *not* in the business of finding jobs for working people. They work for employers and are paid for their ability to find talents to fill slots. It's important to remember this distinction, because it affects every aspect of your dealings with headhunters. *You* are not the headhunter's client; rather, you are the "raw material" the headhunter is "selling" to his or her clients.

In order to fully understand what recruiters can and cannot do for you, it's helpful to know the process that they typically go through when undertaking an assignment. Here's an outline of that process.

1. *Evaluating the company and the business.* The good search consultant wants to know important basic facts about the client employer, such as:

- What is it like to work there? How do the company's culture and history shape the experience of its employees?
- Who runs the company? What are the top management team and the rank-and-file employees like?
- How does the organization work? How do decisions get made?
- How is the company funded? Is it publicly or privately owned? Are venture capitalists or other private funding companies involved in ownership or management?
- How healthy is the company's current financial status?
- What are the company's future prospects? What products or services hold the most promise for its future?

2. *Determining the hiring company's mission.* As a corporate consultant, the recruiter's job is to serve the hiring company in whatever way will

most benefit the company. One recruiter observes that his greatest curiosity always concerns "Where is the client feeling the biggest pain?" The more serious the talent problem the headhunter can solve, the better for the headhunter, because his or her fees are likely to grow in proportion. Thus, the headhunter who is tackling a new assignment will want to ask questions such as the following.

- Where does the company suffer from holes in employee talent?
- What problems currently affect the company's products or services?
- Who are the company's main current competitors? Whom does the company expect to compete with in the future?
- What recent changes in strategic direction has the company undertaken? What internal or external factors prompted those changes?
- What potentially valuable opportunities has the company failed to address?
- What valuable employees has the company lost? Why has this happened?
- What important customers have they recently lost or failed to capture?
- Why is it time to hire someone now?

Note that all of these are good questions for *you* to ask about any company you're considering targeting in your career search.

3. *Writing a position description for each job to be filled.* The position description will include:

- Job title
- Reporting relationships
- Operational responsibilities
- Compensation and benefits offered
- Educational requirements
- Skill and work experience requirements

4. *Crafting a recruitment strategy.* Before launching a talent search, the headhunter develops a targeted description of where the best candidates are likely to be found. This description will probably include:

- The industries where candidates with the necessary skills and experience are likely to be found
- The specific companies in those industries where the best candidates probably work
- The names of the top executives at those companies
- The names of the people reporting to those executives—especially those with outstanding reputations for talent, dedication, and creativity

This description provides a number of starting points for the search process. It's likely that one or more outstanding candidates will be found on the lists of names generated here. If not, they will probably lead to other connections that, in time, will unearth candidates who are both well qualified and potentially interested in the jobs to be filled.

5. *Executing the search.* Of course, this step is the heart of the search process, and every firm has its specialized techniques for finding talented job candidates. But there are many methods that every recruiter will employ. These include:

- Calling sources to ask for candidate suggestions
- Researching data bases for possible candidates
- Reviewing the resumes of individuals who are referred either by the client company or by other sources
- Conducting telephone interviews of candidates whose resumes suggest a possible fit
- Conducting face-to-face interviews with the most promising candidates—primarily to screen out those who won't fit the company's culture

The two or three candidates who emerge from this search will then be introduced to the hiring executives at the client firm, and the winning candidate will likely emerge from that small pool.

6. *Negotiating an employment deal.* The recruiter will serve as the go-between in crafting a deal between the client company and the winning job candidate. Naturally, the headhunter operates here within parameters established by the client company. The salary range, for example, will be set by the company, and can only be exceeded with the company's explicit approval. However, the recruiter will have a strong incentive to consummate a successful "marriage" between candidate and company, because its fee will usually be contingent on that marriage. Thus, the recruiter will do everything possible to create a win-win deal that satisfies the needs of both parties.

THE RECRUITER'S CHALLENGE

The biggest challenge faced by most recruiters derives from the fact that the best employees are usually already working somewhere—often happily so. Thus, the headhunter's job includes not simply identifying a great candidate but also convincing her to leave her current job and join the recruiter's client company. Finding and recruiting truly outstanding candidates isn't easy. Hal Johnson, a senior partner at TMP Executive Search, describes the challenges. "Eighty percent of the people we want aren't looking. I want to represent the top 20 percent of talent for the job. I don't get paid to find average people. I want showstoppers. What are showstoppers? They're people who have worked at 'Tiffany' outfits [i.e., first-class companies]. I know the bosses they've had, I know the thinkers they've worked shoulder to shoulder with, I know twenty-five people who know them and their work. And I know the quality of job, the quality of people, and the resources they've had. Those are showstoppers."

The fees paid to recruiters by hiring companies can be very large, sometimes into the hundreds of thousands. Why are companies willing to pay so much? The answer lies in the intensity of the talent wars, particularly in high-tech business. "People are the only asset a company really has," says Gloree Parker-Roden, senior vice president of Enterprise Services at Pearson Technology Centre. So the search fee may be "a bargain" if the recruiter can identify a really top-flight candidate. "We post jobs on Monster and HotJobs," Parker-Roden explains, "but it takes a lot of interviewing time to go through the resumes we receive and yield good candidates. So it ends up costing about as much as paying search fees to a headhunter in the beginning."

The best headhunters bring both strategic savvy and competitive intensity to the talent wars. "It's like a dartboard," says Mike Nieset, a vice president with Christian & Timbers. "There is a red spot in the middle of the target—the bull's-eye. That is who I am going after. Not the person in the ring outside the bull's-eye. After going through the process, I come up with maybe two or three people on the planet who represent the bull's-eye."

Then the second part of the game—recruitment—begins. "I view it as a team sport," Nieset says. "I get the board of directors involved, and [we] romance the process. We can make it pretty exciting and compelling for an individual to leave a good job and go with the one we have."

"It's not so glamorous," says one headhunter about his work. "Your life is always making phone calls and waiting for people to call you back. And every search is different. Last week, one of my clients was down to the top candidate for an executive job. We prepped the candidate for the interview with the CEO, as we always do. Instead, the CEO had his wife interview the person. It was totally unprofessional, and we lost the candidate because of it. Another search took us more than eight months. We started losing money on it after around five months. We finally got the right candidate and negotiated a great package. We offered her the world. But her original company came back and offered her the world *plus*. So she stayed. That's how it goes sometimes."

RULES OF THE GAME

The recruitment game is not without rules. Headhunters observe a general prohibition against raiding a client company to find an executive for another client. This may not be as simple a matter as it sounds. There have been so many company consolidations in recent years that it can be difficult to tell where one company stops and another company starts. As a result, the traditional prohibition against raiding is sometimes subject to modification. One search firm's main client owns 130 separate operating companies. That would put a lot of people off limits to recruiting. So the agency says, "We'll be happy to put other divisions off limits if we get business there." Another agency only promises not to work for certain competitors.

In addition, headhunters will generally not recruit away anyone they recruited for their current job. Nor will they recruit that person's former boss or direct reports for one year after that person's hire date.

These off-limits policies obviously have an impact on job hunters. If you work for a company that has hired a search firm, the firm won't consider recruiting you for positions elsewhere unless you have publicly made it known that you are leaving, and have notified your current employer.

As you can see, the search process involves quite an intimate connection between the headhunter and the client firm. No wonder some professionals establish long-term relationships with employers, over time bringing dozens of leading managers and executives to the firm and so significantly shaping the culture and business performance of the company.

Types of Headhunters

Not all headhunters are alike. There are basically three different types, and each operates differently. Unfortunately, the names of each type are not always used consistently, so the only sure way to know which type of recruiter you are dealing with is to ask direct, sometimes probing questions. The three categories are:

1. *Employment agencies,* which place lower- and entry-level employees on a contingent basis (that is, for fees paid by the employer after hiring) in jobs paying salaries in the $40,000 to $60,000 range.

2. *Contingency search firms*, which place mid- to higher-level employees in jobs paying salaries in the $40,000 to $120,000 range.
3. *Retainer search firms*, which place higher-level employees in jobs paying more than $120,000 on a fee-for-effort basis.

EMPLOYMENT AGENCIES

Employment agencies are mainly geographically based, focusing on local opportunities. Success in the local agency business is largely a numbers game, based on "who can fax resumes fastest," as one CEO puts it. For each employee hired, the agency receives a fee that generally ranges between 15 and 33 percent of the position's annual cash compensation. Thus, when a $40,000 entry-level job is filled, the agency fee will probably be between $6,000 and $13,000.

CONTINGENCY SEARCH FIRMS

Contingency firms often specialize in specific job levels, categories, or industries. They sometimes have exclusive or near-exclusive relationships with their clients. As with employment agencies, the fees are usually in the 15 to 33 percent range.

Sometimes a contingency firm will latch onto an attractive candidate and shop him or her around among a series of prospective employers. You need to be very aware of this possibility and exercise as much control over the shopping process as possible. Because the agency (naturally enough) is trying to sell you anywhere it possibly can, your resume and your job search may not be handled with discretion. Unfortunately, I know of many situations where confidentiality was promised—and not delivered.

Because contingency firms work with no financial commitment from a client, they tend to operate as freelancers, with little sense of loyalty or perseverance. As one contingency recruiter explained to me, "I assign priority to assignments and work on whatever job I think is going to be easiest to fill. If a search is difficult, I can just drop it, whereas a retainer firm is locked into completing it."

As a result, working with a contingency firm can sometimes prove very frustrating for job hunters. Dale Telford, founder of bITssol, explains these frustrations candidly. "The problem with an agency," he says, "is that they are given a template of what the job requirements are and the background needed. So even though you may be the best at what you do, if you are missing one of the requirements on paper, you will be passed over. Yet if you had the opportunity to meet face-to-face with the same company because someone referred you, the one missing ability would probably be overlooked. My previous job is the perfect example. I had no experience in the technology they were using, but I happened to call an old friend, who is one of the vice presidents, and inquired about possible openings. He put me in touch with the director, and I worked there until starting my own company."

RETAINER SEARCH FIRMS

Retainer search firms work at the most senior level and are paid in advance to fill a specific position, with fees usually ranging between 25 percent and 35 percent of annual cash compensation. Retainer firms also generally get around a third of any sign-on bonus, stock, or stock options. Small, so-called boutique search firms may specialize in a specific industry, while the larger companies (including Heidrick & Struggles, Spencer Stuart, Korn/Ferry, and TMP Executive Search) work in all industries on an international basis.

Many retainer firms offer additional services related to human resources management, such as succession planning, exit interviewing, strategic business planning, human capital auditing, venture capital sourcing, customer retention programs, competitive and market intelligence, compensation surveys, acquisition due diligence and management analysis, and post-merger human-capital review.

As in all industries, different recruiting firms have different personalities and cultures. There are good and bad companies in every arena and in all three categories of firms. One recruiter said to me, "If you lie, cheat, and steal, you want the same type of headhunter. They are out there.

Naturally, if you have strong positive values, you want the same from a headhunter."

What a Headhunter Will—and Will Not—Do for You

Many job hunters think, "I've got to get my resume into the hands of an executive recruiter. They'll find a job for me." That is incorrect thinking. As previously mentioned, the headhunter works for a company client, and his loyalties and concerns lie (or should lie) with the firm that pays his fees. Thus, if you have something special to offer the hiring client, the headhunter will be delighted to work with you. Otherwise, they will be uninterested. "I'm still surprised by the number of people who don't realize [that] we respond to client needs, not individual needs," says Fred Taraschi, managing director of Ken Clark International. "People will send a resume over the transom and ask, 'Please let me know how you'll market me.' I'm not being rude, but we won't."

Furthermore, headhunters generally are not imaginative, far-sighted, or visionary when it comes to matching candidates with positions. To attract the attention of a recruiter, you usually need to have the precise requirements they are looking for. Even if your industry credentials are strong, if you haven't done exactly what they are looking for in a similar company, it's unlikely that a headhunter will place you.

A corollary to this truth is that it's very rare to make a career switch through a recruiter. Headhunters aren't looking for candidates who are significantly "out of the box." They don't want someone who hasn't done it yet—even if the person could. One recruiter told me, "An experienced director may get a shot at a vice president title. And there are times when people's titles will change as a result of a job shift we initiate. But the client company is not usually looking to take on a huge risk. Therefore, we are not looking for a 'cross-pollinator' or a 'contrast candidate.'"

Thus, if you want to drastically change what you've been doing, don't waste your time with professional recruiters. They will not aid in changing your career path. Rather, they will keep you on the track you were on, perhaps with a bump up in level.

What to Do When a Headhunter Calls

If a recruiter calls you, say that you'll call back. Don't take the call right then, even if you have the time to talk. Take a few minutes to think about what you need to find out from them, write some questions down, then call back.

Before getting involved with a headhunter, find out about his or her background in the search business, the industries he or she has worked in, the levels and kinds of positions he or she fills, and the firm's history and experience. It's also okay to ask about where the recruiter got your name.

A typical opening conversation with a headhunter might go something like this.

The headhunter might begin by saying, "We have an opportunity in the technology sector that requires a person with the type of credentials you may possess. From talking to different people, I hear you are the person to talk to," or "We're looking for someone who's in the same place you were in three years ago." Then he or she will give you a three-sentence description of the job.

You might be interested in the position yourself. If so, say so, and summarize your experience background, and career objectives in a way that will sound attractive. If you're not interested, you may want to refer the headhunter to an acquaintance or two. Either way, ask some follow-up questions for your own benefit, such as: What types of companies are increasing or decreasing their search activities? Who are the main competitors of their client company? Why are they looking for the specific credentials they've mentioned? The point is to get some new information to help you, your company, or your friends.

If you handle your headhunter conversations effectively, you can accomplish several things, including making yourself appear more desirable by presenting yourself in a favorable light; helping the recruiter by suggesting some names that might solve his or her immediate problem; and learning something new about your industry and the evolving job marketplace. In turn, the recruiter is going to appreciate and remember you. When the next great job comes along, he or she will come calling again.

The number-one way recruiters find out about people is through their own A-level contacts: people they know who have accomplished notable things in their careers, individuals they've placed before, centers of influence from trade or industry groups, and other trusted friends and colleagues. But creative search consultants use other techniques for tracking down great job candidates as well. Some have been known to offer thousands of dollars to business reporters to learn in advance who will be featured in stories (money that was refused). Others have arranged public speaking engagements for star executives in an effort to win them for top jobs with their clients.

One headhunter explains a favorite technique: "When a candidate I interview speaks in glowing terms of his successes, I know he must have a great team to make that happen. So I find out who reports to him and who was likely responsible for the results. Then I go after them."

Many world-class executives get five to ten calls a week from recruiters. One wireless company president says, "My voicemail is full of job offers. I don't even have time to respond." You may not be in such demand, but if a reputable search consultant leaves you a voicemail message, return it. It's about relationship building—if not for now, for the future. "It's amazing how many people don't practice the basic courtesy of returning phone calls," comments Dave Hardie, a partner with Heidrick & Struggles. "Today the recruiter may just be calling you as a source. In the future, that same recruiter may have a job you'd kill for. But if you didn't return the call when the recruiter originally contacted you, however far back, he or she will remember."

Think of the time and energy you invest in your relationships with headhunters as part of your long-term career strategy. "Keep an active network of a handful of recruiters whom you know, respect, and trust," says Katherine Cizynski, a senior partner with Wiser Partners. "It will create a safety net if you suddenly find yourself in a situation where you are looking for a job, whether by choice or unexpectedly. It also gives you a sounding board for career advice. As a quid pro quo, you should readily make yourself available to these recruiters, providing referrals and references, perhaps even bringing them in to handle a search on your behalf."

Contacting a Headhunter Directly

As with virtually all kinds of career contacts, the best way to introduce yourself to a headhunter is through a referral or recommendation of some kind. But if you must make contact without a reference, submit your resume via e-mail, fax, or mail. Attach a cover letter that is as specifically customized to the recruiter as possible, referring to their specialty, background, and expertise. Follow up with a call to confirm they've received it.

Here's how a headhunter typically handles an unsolicited resume: The front page goes into a file. A thank you/acknowledgment letter goes out. The resume is then read thoroughly for entry into the headhunter's database. Your school and its location are listed (because people frequently end up working within sixty miles from where they went to school), followed by your graduation date and major. Then your job history is detailed: who you worked for, what you did, when and where you were employed. Finally, your current career objective is listed.

If you get a chance to talk with the recruiter, be prepared with your two- or three-sentence summary of what you're looking for; for example, "I'm looking for a job as vice president of finance at a start-up, preferably one that has gone through a couple of rounds of financing already. I don't want to join a company that consists of five guys with an idea and a dog. Are you working on anything like that in my area?" The recruiter will generally give you an honest answer. If it's not what you want to hear, ask whether the recruiter can recommend someone else for you to talk to.

If the headhunter isn't handling any positions for which you're a fit, you might try this follow-up question: "Since you know people at ABC company, could you tell me who I might contact on my own?" The question flatters the headhunter by acknowledging that he or she is on top of what's happening. And what's the worst thing the headhunter can say if you ask for company information, a name, or a referral? "No." Surprisingly few headhunters will refuse to help. They don't want to appear uninformed, out of touch, out of the loop, or indifferent. Remember, they always have something to gain by helping you. You are, at the very least, a prospective corporate client for them in the future.

Don't blindly send a mass mailing to executive recruiters. It will do nothing for you except make you feel productive. And don't expect miracles from the headhunter contacts you do make. Among the many complaints I hear from job hunters about recruiters are: "They never call back," "I could have done the job they were looking for," and "They act like they couldn't care less."

These complaints have a modicum of truth to them. Many headhunters don't call after receiving your resume because they receive hundreds or thousands of them. But it's not true that headhunters routinely ignore qualified candidates. As one recruiter puts it, "If I thought you could do the job and that you would make my client happy, I'd be all over you." Remember, the headhunter is in business to solve his or her client's needs—not yours. Unless you fit the exact job specifications, they'll pass over you in a flash.

Use headhunters as sources of information, contacts, referrals, and ideas, but don't expect them to work on your behalf. In the job market, you are your own best representative.

Some Truths to Remember About Headhunters

In the numerous meetings I've attended with executive search consultants, I've observed that headhunters are quick to label people. Candidates who are up for discussion get tagged within a few seconds as either "on target," "near target," or "out of the box." A job hunter may spend weeks or months on resume preparation and company research, only to have it all go up in smoke with one comment from a recruiter.

Here's a sampling of some of the devastating comments I've heard in headhunters' meetings—comments that broke the chances of particular candidates.

- "Alan's not an A player."
- "The reference said Ray is an operator but has no vision."
- "The company Bob is with doesn't have a reputation for being well managed."
- "Keith's a good guy, but there's no technology background in his career."
- "We're looking for someone with more pull than Allie."
- "John is too slick; you never know what he's thinking."
- "We want someone with broader industry experience."
- "Looks good, smells bad."
- "If you asked Jerry for a reference, you'd get twenty bad ones."
- "I'm a little hesitant about Joe. He's heir apparent [at his current company], and he's only had experience at one company. I'm afraid he won't scale."
- "Ben's wife is a decorator who was doing work for a board member. She told him he should hire her husband, and that's how he got the job."
- "He's not a leader."
- "She's not a consensus builder."
- "He has trouble networking."
- "He was at the battle, but he sat behind the lines."
- "She's a loose cannon; they call her Wacky Jacky."
- "He's got one real flat spot on the wheel."
- "He's solid, but not world class."

Needless to say, you wouldn't want to be on the receiving end of any of these or similar comments. On the other hand, comments like the following can push you to the head of the pack.

- "We think he's a good guy."
- "Makes the best presentation I've ever seen."
- "Runs a crisp organization."
- "Mark manages problems in a nondisruptive way."
- "Tim wants this job; he's looking to hit one out of the park."
- "If you could get Karen on your board, that would be very smart."
- "You might as well forget Phil—you'll never get him."
- "He'll take the hill and hold it."
- "He's totally jazzed and motivated, with no ego."
- "Janice builds trust."
- "He's a go-getter and a real team player with great personality."
- "Paula's a water-walker."
- "Pete is world class."

It's sobering to realize how one comment from the right person can change your life. One search consultant told me, "We're right about 75 percent of the time." With the weight their words carry, they should be right 100 percent of the time!

Always remember that the headhunter is not your friend. The relationship may at times feel like a friendship. You may talk with a recruiter thirty or forty times during the search process. He or she may meet your spouse and talk with your children, and you may feel that a great rapport has developed between you. But you must maintain a degree of separation. Don't air your dirty laundry, confess your mistakes, expound on your misgivings about the job, or rely on the headhunter's advice.

One recruiter tells me that candidates frequently ask him, "Should I ask for more money?" These candidates are forgetting that the headhunter represents the hiring company. "There are times I have the latitude to offer another twenty grand if that's what it's going to take to get a candidate off the fence. But if I don't have to give it, I'm not going to, because I represent the client's interest."

Again, the crucial truth to remember is a simple one: *The headhunter is not your friend.*

Venture Capitalists as Job Sources

Venture capitalists (VCs) have enormous influence in the talent business because of the financial power they wield. With 1,700 companies receiving $24.6 billion in venture capital funding *in one quarter* last year, it's clear that venture capitalists are a force to be reckoned with in the job market.

The venture capital business has been around only since the mid-1970s. Venture capitalists are catalysts, enablers, and cheerleaders, though not creators of companies. They have influence in the industry, know people with contacts, know how to attract the best executive talent, and take their companies public faster than others.

A successful VC has the ability to spot a deal that's going to work. Think of thirty-three-year-old Steve Jurvetson of Draper Fisher Jurvetson (DFJ), who invested $300,000 in Sabeer Bhatia's Hotmail after it had been turned down by twenty-one other VCs. Hotmail was later purchased by Microsoft for $400 million, earning DFJ $100 million. Then there was DFJ's $700,000 investment in Kana Communications, now valued at $1 billion. And DFJ isn't even considered one of the big guns in the VC world. Those include Kleiner, Perkins, Caufield, & Byers, which funded Amazon.com, Sun Microsytems, and Netscape; Sequoia Capital, which funded Yahoo; Benchmark Capital, which funded eBay and WebVan; and the Mayfield Fund, which backed Silicon Graphics, 3Com, and Genentech.

A typical five-person VC firm may review 5,000 business plans a year and invest in ten. Out of those ten, five may be total write-offs, three will be modest successes, one will earn double the initial investment, and one will return 50 to 100 times the investment. Naturally the goal is to sign a number of big winners and only a few losers.

In reviewing a business for possible funding, a VC firm looks for such qualities as the following.

- A successful track record in anything
- Strong management
- Attractive alliances with other companies
- Products or services that offer clear, unique benefits to the customer
- Real revenues and earnings and healthy gross margins
- A compelling business plan on which a sustainable long-term business can be built

These are qualities you can also look for in companies as you conduct your job search. Jack Shields, a general partner with Best Capital Ventures, sums up what VCs are looking for this way: "We see exciting, interesting technology every day of the week. Its uniqueness is important. But the most important aspect [of any business] is the management team. Technology comes and goes. An average team with superb technology will not do as well as a superb team with average technology. The people make it work." Thus, just as a VC firm looks for a company with a strong team, that should be the kind of company you seek to be a part of.

Venture capitalists generally retain recruiters—lots of recruiters. As of this writing, VC-backed companies have six hundred openings for chief executives alone, so you can see the scope of the need.

Some VCs review resumes that come in. If a person seems appropriate for a job at one of the VC's portfolio companies, the resume will be forwarded. "We do proactive candidate development," says Norbert Gottenberg, a recruiting partner with Kleiner Perkins. "We target companies, and we target individuals in them that we should know. Then we develop a meet-and-greet strategy so we know the players. We're always on the prowl for board-level executives."

Venture capitalists don't have the time to do most search work themselves. However, they do have a lot of influence, as Kathleen Ho O'Donnell, a principal at VC firm Advent International, explains:

> **VCs get involved themselves in the hiring of [senior] executives, so contacting VCs directly can be effective. For example, if one**

> of my companies is planning to hire a senior exec (a CEO, COO, CFO, or VP of sales, for example), I will be one of the first to know and I will be on the lookout for good candidates. If I happen to get a good resume, I'll pass it along. As a director and investor, my opinion will carry a lot of weight, and my support of a candidate will be required. So anyone I pass along will at least get an interview. Of course, I will also recommend a headhunter that I know, unless I already know of an excellent candidate that I feel precludes a search.
>
> For nonsenior management positions—anything from an administrative assistant to a sales person to a mid-level manager—VCs don't get directly involved. But they do pass along very qualified and highly recommended candidates to their portfolio companies—usually friends or friends of friends. For example, if a friend of mine is looking for a job in a marketing department and I think highly of her, I will pass her resume to the partners in my firm and to my own portfolio companies, asking if there are appropriate positions available. My own portfolio companies are usually very quick to get back to me, because I am a director and investor in the company. This is a good way to get an initial interview, but it is very hit-or-miss, because there may very well be no appropriate positions at the portfolio companies.
>
> Something I have seen people doing more frequently is searching the press for exciting companies, going to their Web sites to see who the investors are, and then searching their personal networks for contacts at those investment firms. VCs are *very* busy people, so they typically respond only to requests from friends or partners within the firm."

If you're interested in getting onto the radar screen of a VC firm, start by doing your homework. As one of the partner's assistants at Centennial Fund says, "You can get on our Web site, see who we've co-invested with. It's not rocket science. There's just a little digging required to get information in advance."

Another partner's assistant emphasizes the importance of a referral: "We get so many resumes, and they all look good. Each one is dazzling, and it's impossible to determine who's really great and who's not. For that reason, it's best to have someone the partner knows send the resume."

Incubators as Job Sources

An additional source for job opportunities is an *incubator*—an investor who offers a start-up company everything from furnished office space, legal advice, executive recruiting, human resources and accounting services, and salary and bonus data, to advice on loans and real estate, marketing, debt financing, public relations, as well as contacts from the incubator's "golden rolodex"—all with the goal of getting the company up and running quickly. Two of the best known incubators in the high-tech world are Idealab and Safeguard Scientific.

With incubators, it's all about speed. Mohanbir Sawhney, professor of electronic commerce at the Kellogg Graduate School of Management at Northwestern University, likens working with an incubator to an Indy pit stop: "The car comes in, and—bang, bang, bang—twenty guys work on it, and they're off in thirty seconds." Most companies that work with incubators stay with them for about a year.

Finally, be aware that a single company may have a relationship with an incubator, with several so-called *angel investors* (individuals who invest heavily in companies), and multiple VCs. If you're considering joining a start-up, dig into its background and try to determine the identity of *all* its owners. The more numerous they are, the more masters you may have to answer to, and the less stock there will be available for current and future employees.

If you choose to use an incubator as a job source, your chances for success will be greater if someone connected to the firm refers you. Further, handle your relationship with the incubator as you would a VC firm. Remember that the firm's primary goal is the success of the company it is working with. Thus, don't expect your personal interests to be a top concern for them.

8

SELLING YOURSELF: THE RESUME AND THE INTERVIEW

"First, make sure you are compatible with the job in attitude, aptitude, competency, prowess, skill, acumen. Second, think and behave like an entrepreneur. Third, give feedback—and seek it also. Fourth, think team. Fifth, solve problems. And sixth, display emotional intelligence. Know how to relate to other people and their aspirations. Because in life, you have to get things done with people or through people."

—Ross Brown, vice president for human resources, Analog Devices

Tools for Job Hunting

You may have been surprised to find that, unlike many books on career development, this book does not begin with information on resumes, cover letters, and job interviews.

There's a good reason for that. If you've done a good job on the compass-setting exercise described in Chapter 3, your resume is largely written. At the very least, you've gathered an enormous amount of important, interesting, and relevant information about yourself, your skills, your experiences, and your aspirations, all of which will make crafting the resume far easier than it is for most job hunters. And if you've worked on becoming known using the networking techniques explained in Chapter 5, you've already been developing and practicing most of the skills needed to give a great job interview. In effect—whether you realized it or not—you've done the most important groundwork for the resume and interview already.

There's another reason why this book doesn't start with the resume and the interview. Too many job hunters focus on these tools, often to the exclusion of more important strategies. Both are important, of course, but neither the resume not the interview is the be-all and end-all of the job search. Neither one, by itself, will win you your next great job, and neither one deserves the kind of obsessive attention many job hunters give these tools. Above all, keep the resume in perspective. Don't belabor either one. Some people endlessly work and rework their resumes, in search of perfection. The truth is that there's no such thing. Go with 85 percent of perfect, and you'll do just fine.

Crafting Your Resume

To begin developing your resume, pull out the notes from your kitchen-table compass-setting session. Review that information and relate it to these questions.

- Who is the audience for my resume? In other words, what specific companies or industry am I seeking entry to?
- What qualities will the hiring managers in those companies be looking for?
- What have past employers liked about me that might cause a new employer to want to hire me?
- What do I want a potential employer to know about me?

With these questions in mind, go through your notes. Highlight facts and details you want to include on the resume. These may include career accomplishments, educational achievements, skills and talents you've developed, awards you've received, and so on.

How long should your resume be? There's no one answer. After you've written a first draft, revise it to cut out excess verbiage. Edit out filler stuff, including unnecessary jargon. If you've done that and what remains is three pages long, that's okay. If it's a poorly written resume, even one page will seem too long. "The more information, the better," says Michelle Monfor, an executive recruiter with TRRG. "Don't be shy. Give lots of information about your responsibilities and accomplishments. When they say your resume should never exceed two pages, that's wrong advice. If you have more to say, say it."

Parts of the Resume

The standard resume includes the following parts. This explanation will help you determine where each of your highlights should appear.

- **Name and contact information.** Top off your resume with your full name and all your relevant contact information: postal address, phone and fax numbers, e-mail addresses. Include all the point of contacts that you are comfortable using for your job search. *Don't* use a phone number or e-mail address from your current employer; it's tacky and unprofessional to appear to be job hunting on company time.

- **Job objective.** This is a brief (two- or three-sentence) description of the kind of position you want. Don't leave the reader to do the job of trying to discern your interests or goals. Be clear and specific about what you want. Who knows? You may get it.

Your job objective should also show the value you'll bring to your next employer. Don't just write about what you want in a job. Indicate what you can do for them, especially in terms of bottom-line impact.

• **Qualifications summary.** Some job hunters choose to follow the job objective with a snapshot of their leading qualifications. If you go this route, make sure the summary highlights accomplishments and skills that match the interests of a prospective employers. Title the section "Summary," and in bullet format list items such as the following:

- The number of people you've supervised
- The dollar amount of budgets you've controlled
- The range of job titles you've held
- Increases in sales, profits, or productivity that you've spearheaded
- New products you've developed or launched
- Programming languages or foreign languages you've mastered

These bullets should be brief, just five to ten words each. The point is to give a quick overview of the information your readers will see in depth as they read on. Most recruiters invest a minute or less in the typical resume. A well-written summary can be an effective way of capturing a recruiter's interest within a few seconds and persuading him or her to devote a little extra time to your story.

• **Accomplishment case stories.** This is another section you may want to include early in your resume. Here you provide two to four brief case stories, each with a title such as "People Skills," "Teamwork," "Leadership," "Negotiation," or "Organization," denoting the specific business talent the story highlights. Then describe the situation or challenge, explain what you did, and the concrete results you obtained. Each case story should run just twenty to forty words in length.

In her job search, Jana Weaver, whose story you read at the end of Chapter 3, included both a qualifications summary and a set of accomplishment case stories. She credits these with capturing the positive attention of prospective employers. Jana says, "My new boss says he saw my resume and told the recruiter to get me in as quickly as possible."

Some recruiters like the case story method; others don't. Some readers find it confusing because it violates the traditional chronological format that most resumes follow. You may want to try having one resume with case stories and one without, and test the varying results for yourself. If you do use case stories, be sure to articulate your bottom-line contributions using quantifiable results rather than puffery or vague self-praise, which tend to produce a sense of suspicion or mistrust on the part of recruiters. Mike Nieset, a vice president with Christian & Timbers, takes a jaundiced view of most case stories: "They are usually the job hunter's own embellishment of what they did. I want to know literally what happened instead of what they think or say happened. . . . I want facts."

- **Work history/professional experience.** This is the traditional heart of the resume, where your career story is laid out in reverse chronological form (that is, beginning with your current or most recent job and working backward). Provide the company name and include a brief description, even if it's a well-known company such as IBM. If it is a smaller, less well known company, briefly explain the company's current status, size, and the product or service provided (for example, "A privately held manufacturer of specialized computer networking gear with annual revenues of $400 million"). Also indicate the division or department you worked for, its size, and your job location.

It's also important to show (and be able to explain in an interview) your reporting status, both upward and downward. A particular job title may suggest far more responsibility and significance in one company than in another. "For example," as Mike Nieset explains, "someone with a general manager title at one company may have just sales [people] reporting to him, whereas at another company [someone with the same title may have] a full cross-functional team reporting to him. I want to know that. Or, if you're a CFO and you *don't* report directly to the CEO, I want to know that."

For each job listed, include your title(s), dates of employment (in months and years), and a brief description, in bulleted form, of your duties and tasks. Weigh your current and recent jobs more heavily than jobs from five or more years in the past. Include crucial technical keywords for your field or for the field you are hoping to work in (for example, *Unix,*

HTML, database scaling, shell scripting, Windows NT, and so on). Including appropriate keywords will help ensure that a computerized scanning program will spot your resume.

Make sure your bullet points are as specific, clear, concise, and objective as possible. For example, instead of the following, rather vague explanation: "Led my team in development and execution of an operational plan that achieved significant success for the division," rewrite the bullet to provide much more specific detail about the success achieved, as follows: "Led a six-member team in development and execution of a plan that surpassed sales quotas by at least 30 percent in each of four successive periods."

You want the recruiter who reads your resume to see that you are a creative thinker, a decision-maker, and a problem-solver—someone who can handle responsibility, has healthy self-esteem, has strong people skills, and is honest and ethical. You can convey this kind of impression by crafting your bullet points to include *action words* such as the following.

- Increased [sales or profits]
- Motivated
- Reduced [costs or errors]
- Conceived
- Completed
- Presented
- Negotiated
- Solved
- Accelerated
- Tested
- Recruited
- Controlled
- Created
- Executed
- Developed
- Established
- Directed
- Expanded
- Improved

Such action words offer a hard-hitting, succinct idea of what you can do for your next employer. Of course, you should be prepared to back up and elaborate on any results or accomplishments you claim with a detailed explanation when you're interviewed. Be ready to answer the obvious "who, what, when, where, how, and so what" questions. Also be prepared to explain how and why you left each job as well as any gaps in employment dates.

Be sure to write the job bullets to highlight your skills, suggesting how they can be applied to your next job. This is particularly critical if you are in a job or career transition. "You need to generalize your resume to the work you can do, not get trapped in the industry that you were in," says Amy Duclos, president of Colorado Corporate Search LLC. "In our searches, we're looking for the right attitude, not just the right history." Write your resume to convey your winning attitude, not just the bare facts of what you've done.

- **Educational background.** This section should include both college and graduate-school education and all forms of continuing education—on-the-job training, seminars and workshops, distance learning, night and weekend classes, and so on. If your formal schooling was both stellar and recent, you may want to insert this section *before* the work history. This applies especially to brand-new or recent grads and to those who've recently completed training in preparation for a new career or industry. Thus, if you just completed a degree in programming to support a shift from (say) the real-estate business, you should list your programming degree ahead of your experience as a Realtor.

There will probably always be a bias toward graduates from prestigious schools, but in the high-tech world, the most prestigious schools aren't necessarily the ones you'd assume. Paola Bonomo, a partner at McKinsey & Company, is most intrigued by those who attended schools that are considered "hotbeds of innovation," such as Stanford, MIT, Helsinki University of Technology, and Technion in Israel. Time spent at such places is valuable, according to Bonomo, "not just because of what you're taught but because of the air you breathe and the myriad of ideas you are exposed to."

A recent *Fortune* magazine survey found that, of the forty wealthiest *young* business people (younger than forty):

- Four graduated from Ivy League schools.
- Four graduated from MIT.

- Three graduated from Berkeley.
- Three graduated from Stanford.
- Twenty graduated from other schools.
- Six were college dropouts.

Some in the high-tech world are even biased *against* Ivy Leaguers. One VC secretary told me, "When we see someone flaunt 'Harvard MBA,' it sort of puts us off. To us, what you've done is more important than where you went to school fifteen years ago."

• **Additional information.** This category of data, sometimes headed *Personal,* comes last on your resume. It should be brief. Include an interesting fact or two about your activities outside of work to show that you are a multidimensional, balanced person. Here's where you'd include leadership positions in your community, awards received, articles published, athletic activities, and so on.

Mary Petryzian, director of engineering for Raytheon (Rocky Mountain Region), says that she looks for community activities when reading a resume. She thinks it makes for a better employee. Jerome Davis, president of Business Process Management at Electronic Data Systems, mentions in his resume the fact that he played in two Rose Bowl football games. Is this directly relevant to his career? Maybe not. But it reflects his competitive spirit, discipline, and determination—and it helps to get his resume noticed and remembered.

Honing Your Resume

The following resume-preparation guidelines will help you craft an effective resume.

• Once you've listed all the information you want to include on your resume, organize your notes into a streamlined, graphically interesting format. *Graphically interesting* doesn't mean fancy or affected. A judicious use of bold face, italics, underlining, bullet points, and indention will suffice to make your resume clear and inviting to read. Further, be consistent. For example, if you decide to italicize one job title, italicize them all. This will make it easy for a reader to comprehend your resume at a glance.

• Use only one or two type styles, in a readable size. For the body of the text, a serif font like Times Roman is better than a sans serif font. (You can recognize serif fonts by the little cross lines at the top and bottom of most letters; sans serif faces don't have the cross lines.) A sans serif font is all right for headings or for your name and address at the top of the resume. And don't use small type. Depending on the particular font you choose, twelve-point type is usually ideal, while ten points or less is almost always too small. Leave room in the margins for recruiters to write notes or questions.

• Print your resume on good quality white or very pale cream, blue, gray, or green paper—nothing neon-colored or bold. Avoid papers with preprinted margins, background designs, or other embellishments. They take attention away from you and your background. (I've actually heard some misguided job hunters say that they use fuchsia-colored paper or print their address upside down "to draw attention" to their resumes. It probably works—the resume draws attention on its way to the trash can!)

• Be meticulous about correcting typos, spelling errors, grammar mistakes, and other errors. Don't rely on your computer's spell check feature—it can't check proper names (such as company names), nor can it recognize a misused word (such as their for there). "When there are typos [on a resume]," says one headhunter, "I figure the person either isn't smart or has spent no time thinking about what he or she is sending out."

• Don't pay a professional to write your resume for you. You won't be as intimately familiar with it as you would be if you wrote it yourself, and it won't reflect your story and your personality as vividly as it should. If the services of a professional writer are available and paid for (for example, if you've been provided with resume-writing services as part of an outplacement package), write your own resume at the same time the professional is drafting his or her version. Then compare the two and combine the best elements from both versions in a final resume.

• After drafting your resume, set it aside for a few days and read it again with a fresh perspective—as if you were a recruiter, not the author of the resume. Have a friend or two read it and comment as well.

• Keep your resume up to date, even when you're not actively in the job market. Bill Manning, a senior electrical engineer with NASA, says, "I always have my updated resume circulated, even if I don't want to change jobs. I like to know what my market value is at all times. . . . You never know when the ax will come. There are budget cuts all the time. . . . It's also flattering to be offered great jobs. And the best time to find a better job is when you have one."

Play It Straight

Whatever you do, don't lie, stretch, or spin the truth on your resume. If you do, you may get away with it for a while, but most industries—including high-tech industries—are small worlds, and the chances are good that one day the truth will catch up with you. When this happens, your reputation will be seriously damaged, and so will your career.

Jennifer Hopkins, vice president and general manager with GSU/Agilent Technologies, tells about a resume she received. "The individual sounded too good; the numbers on the resume were a little too big. I checked [and found that] he embellished dramatically, [claiming budget responsibility of] something like $400 million instead of $20 million." You can guess what happened to that candidate.

When you are close to receiving a job offer, especially at the executive level, you may be asked to sign a consent form for a background check. With or without explicit consent, most employers do such checks on prospective employees, especially now that technology is making it easier and faster than ever. They contact previous employers, verify educational claims, run criminal conviction and civil case searches, and may even check your credit history—all electronically. Most of the data is a matter of public record, and some consultant services can provide companies with background searches about job hunters for as little as fifty dollars per candidate.

Perhaps you're thinking that a savvy job hunter might be able to cover his or her tracks and make it hard for a prospective employer to do a background check. Even if this is possible, it'll probably be counter-productive. "Make it easy for your history to be checked on," one CEO advises. "I had to hire two private investigators to find out about one guy before I was ready to hire him. Making it difficult makes it look like you have something to hide."

If you have something odd in your background, bring it up at the right time in the interview. It's far better for you to tell a recruiter about a red flag before it turns up unexpectedly.

Other Job-Hunting Documents

Aside from a resume, there are several other documents the savvy job hunter can use. They include the following.

- **A bio.** A resume tends to get dispersed, posted, copied, and forwarded, so you never know in whose hands it will end up. That's why, if you aren't job hunting on an active, public basis, you may want to have a bio or fact sheet about yourself rather than a true resume. This is especially popular at the higher corporate levels, where having a resume would make it appear as if you are job hunting. The preferable perception is that "a person of your caliber" would never have to do that.

The same information is included in a bio or fact sheet, in a less-formal format than in a resume. The bio reads like a several-paragraph narrative describing your background, career highlights, and achievements, without a lot of dates, numbers, or other specific details like those featured on a resume.

The single-page bio would usually have your name, title, company, and work contact information at the top. The first paragraph should describe your current responsibilities and some of your main accomplishments in your present job. The next one to three paragraphs should explain where you've been, including past titles, job responsibilities, and a summary of key initiatives you spearheaded.

A final paragraph should describe some of your nonjob-related experiences, including awards, committees, boards, and memberships that support your professional competence. The bio ends with a one-sentence summary of your educational background.

There's nothing flashy, fancy, or sales-oriented about the bio. Instead, it strives to be interesting, informative, and brief. It's a particularly good tool when you are out getting known, but aren't ready yet to "turn up the juice" with a full-fledged job campaign.

• **The cover letter.** The cover letter goes with the resume (or bio). Its purpose is to connect with the recruiter or other recipient, explain why the resume is being sent, and capture the reader's interest.

Generally speaking, every cover letter needs to be customized for the particular recipient. However, some basic elements can remain the same from one letter to the next.

Keep every cover letter short and simple—four brief paragraphs is ideal. First, mention the person who referred you, then note something about that person and the reason you are writing. Third, explain how your background supports the previous paragraph. Fourth, conclude with a suggestion that you will be happy to meet with the recipient at any mutually convenient time.

If you send a resume and a letter, don't say the same things in both documents. They should complement one another, not duplicate. If the resume is well done, you may need only two or three sentences in your cover letter.

• **Other written materials.** Your career campaign can be enhanced by the preparation of additional written materials. These could include:

- Accomplishment case stories (as described in the resume section earlier)
- Quotes or testimonials about your work from significant people
- Excerpts from articles about you or interviews with you
- Excerpts from articles or reports you authored

- Relevant images, such as photos, awards, citations, and other honors

It would usually be inappropriate to send such materials along with a resume. However, you may want to gather them into a portfolio that you bring along on an interview, to share as appropriate. A personal Web site is another excellent way to make this kind of material available to anyone who is curious about you.

Of course, the purpose of all these materials, from the resume to the cover letter to the bio to the Web site, is to interest and impress potential employers so they'll want to meet you. Your job-hunting documents will have done their work if they win you an interview. When that happens, your job campaign shifts into a new mode, one that requires all your personal connection and communication skills.

Handling the Interview

Before going on any interview, become informed about the company, the industry, and the job. Don't expect your interviewer to educate you. If you haven't done your homework, figured out how big the company is, learned about its products and services, and have no idea where you might fit in, you're wasting your and the recruiter's time.

"Don't call me and offer yourself as the solution to world problems without knowing my company and our situation," says one company executive. Another—Ron Dobson, president of ADP Canada—remarks, "It's amazing how many folks I interview don't know anything about my company, didn't read the annual report or visit our Web site." Dobson urges, "Don't come to an interview without any knowledge of the company, saying, 'Here are my credentials, what do you want me to do?' with no target at all."

Walk into the interview with a clear idea of what the hiring company does, how it earns its profits, and how your talents could help the company do even better in the future. Jeff Irby, a managing director with KPMG Consulting, puts it this way: "I want the person seeking the job to articulate how his or her personal strengths will help my company with its core mission. For example, we are in the consulting business in the financial services industry. Candidates who can demonstrate that they understand the business model of consulting, with its challenges and joys, coupled with how they would apply their skills to help customers, really impresses me. It rarely happens, however."

On the other hand, don't study the company in hopes that you can twist and spin your background in order to fit precisely into a niche within the company. The truth is that, as a job hunter, you can never know precisely what any particular company is looking for. (Even the hiring company is often a little vague about what they really want.) The only solution to this dilemma is a simple one: Don't try to guess what the employer is after. Instead, just be yourself, without spin. If the fit between you and the company is a good one, that will become obvious in the course of the interview.

One executive puts it this way: "When I was in a position to hire someone to work for me, most of the folks tried to cater to what they thought I wanted to see. After hiring these people, I found out that some things that had been stated were simply not true but had been used to get in. We both ended up losing." Don't play the game that way.

EFFECTIVE COMMUNICATION

Another interview reminder that should be obvious but isn't is *Listen.* A productive interview is a two-way conversation, not merely an interrogation or an audition. If you spend the entire interview either talking or waiting impatiently to talk again, you'll learn nothing—and you'll probably make the interviewer uncomfortable as well. Only by listening can you be and appear truly responsive to the interests and needs of the interviewer.

Many job hunters wonder how to handle touchy or uncomfortable subjects, such as the time they were fired, their less-than-friendly relationship with a past boss, or the financial troubles of their current employer. It's important to prepare in advance for these topics. Think through what you want to say, even planning specific phrases to use to describe what happened in a professional, honest, but tactful fashion. Avoid criticizing or bad-mouthing former or current employers or colleagues. And try being the first to bring up any uncomfortable topics. This demonstrates your self-confidence, lets you deal with the issue in the way you want, and gets it out of the way so you can stop worrying about it.

EFFECTIVE SELF-PRAISE

Some job hunters feel awkward or embarrassed speaking about themselves—especially when it comes to saying good things. Many people have been raised to be modest, which makes self-praise feel like "bragging" or "showing off." If you're in this category, here is some advice. First, look for objective, factual statements you can make about your work and accomplishments. Instead of saying, "I'm a great manager," you can say, "During the last three years, my department increased productivity by. . . ." This will feel less boastful, and it will also be more convincing to the interviewer.

Second, try attributing positive statements about your work to others rather than to yourself. Instead of saying "I'm good at . . . ," say, "My peers [or colleagues, boss, or whoever] have said I'm good at. . . ." This way you can get the point across without sounding as if you're bragging.

Third, give others the credit they deserve for accomplishments in which you share. For example, say "I've been fortunate to work with a great group of programming specialists. Last year, we were able to achieve. . . ." This takes away any boastful tone and shows that you are a team player rather than an egotist.

Perhaps the most useful piece of interview advice is simply this: Be succinct. Too many nervous job hunters start to ramble. If this happens to you, a kind-hearted interviewer will try to save you from yourself by interrupting and redirecting you. Others will just let you drown. Avoid the problem by answering the question, then shutting up. If the silence that follows makes you tense, try asking, "Does that answer your question?" or "Is there anything more I can tell you about that topic?" This gracefully puts the ball back in the interviewer's court.

Classic Interview Questions to Prepare For

One aspect of interviewing that makes many job hunters nervous is having to answer questions about themselves. Again, thoughtful preparation is the key.

It may sound obvious, but make a conscious effort to answer the question that the interviewer asks—not questions he or she didn't ask, which could raise issues you shouldn't or needn't discuss. Interviewing is not confession. It's an exchange of information that both parties need to know. Don't volunteer superfluous information that can derail the process. Otherwise, you may be using your valuable face-to-face time telling the interviewer things he or she doesn't want to know.

Most of your answers should run no more than a minute in length. If this sounds very short, pull out your watch and time a minute right now. It's a surprisingly long time, isn't it? Chances are good that, if you can't explain what you've done on the job in sixty seconds or less, you don't really know what you've done. So practice being succinct.

Also practice answering questions using examples, stories, and analogies to illustrate your points. People understand, believe, and remember ideas that are stated through stories much better than ideas expressed in abstract terms.

What follows is a lengthy list of typical interview questions. Although you don't want to sound overrehearsed when you are interviewed, the first time you answer an important career question shouldn't be in an interview. You need to have "tried on" your answer a few times to make sure it fits. So take time before your first big interview to practice answering the following questions, whether you do it to yourself, to a tape recorder, to your roommate or spouse, or to a career mentor. Chances are good that you'll encounter at least a few of these—and variations on them—more than once during the months and years to come.

- Tell me about yourself.
- What do you consider the single most important event in your life? Why?

- Tell me about your current [or last] company.
- Which of your jobs have you liked the most [or the least]? Why?
- What do you know about our company?
- What interests you most about the job you're interviewing for? What interests you the least? Why?
- What do you see as the most important trends in our industry? Why?
- Why did you join your current [or last] company?
- What's the most important aspect of your current [or last] job? Why?
- What type of individuals do you have trouble getting along with? Why?
- Why did you make that job [or career] change?
- Describe two or three major accomplishments so far in your career.
- Why are you considering leaving your current job?
- Why are you interested in this position?
- If you aren't offered this job, what will you do instead?
- What do you see yourself doing five [or ten] years from now?
- How do you make important decisions on the job? Please describe a typical example, and tell me how that decision worked out.
- What are your major strengths and weaknesses?
- What was your worst on-the-job mistake? What did you learn from it?
- If you're hired here, what would you do your first day on the job?
- How do you feel about your current salary? Why?
- Do your colleagues like you?
- Have you ever fired someone? How did you handle it?
- Have you ever been fired? Why?
- What are your goals in life?

Had enough yet? If not, here are some very probing questions used by the executive search firm of Russell Reynolds (as revealed in *Fast Company* magazine).

- When was the last time you had a breakthrough in business?
- When was the last time you took a risk and failed?
- Describe a time when your business wasn't meeting the expectations that you had set out to meet. What did you do to change that?
- What's important to you and to your business?
- How do you define your success? Your team's success?
- How do you communicate your company's vision to the team?
- How do you use resources?
- How do you figure out what your customers want?
- When was the last time that you drew from your experience and insights to fill in a blank when faced with incomplete information?
- Are you more interested in getting stuff done or playing by the rules of the organization chart?
- What does an ideal organization look like?
- How do you use the talents of your employees?
- How do you plan for future growth?
- When was the last time that you changed a tightly held view?
- Do you love to put yourself in the middle of feedback groups?
- When was the last time that you learned something that led to an "Aha!" moment?
- Do you know what it takes to ruthlessly prioritize?
- Can you zoom out to see the most important objectives, instead of allowing yourself to be distracted by tangential tasks with little payoff?

Every job interviewer has his or her unique methods, and some companies pride themselves on their distinctive interviewing techniques. Interviewers at Microsoft, for example, like to ask various kinds of intellectual brainteasers: "If you had a plane full of water and had to land, what would you do?" or "Why are manhole covers round?" or "How many piano tuners are there in the world?" The point is not to come up with the "right" answer but rather to demonstrate intelligence and creativity in thinking through the question.

Markus Schweig, a manager at Microsoft, adds, "I don't ask questions that allow a yes or no answer. I want people to give a scenario they've lived through, and have them tell me how they were more effective than the next [person]. . . . I also like to arrange for an interview with the person's future partners inside and outside of the company—people the person will be working with, even in other departments—and get their opinion [of the candidate]."

One CEO said he asks tough questions to see how the candidate handles stress. Another recruiter comments, "I listen to what they say, then drill down: 'You did this, then what happened?' I dig. I get the history, then go smaller and smaller." Still another says, "I ask questions I already know the answer to, to see how honest they are."

What Are They Looking For?

Behind all the specific interview questions and techniques are a set of more fundamental issues—basic facts about you that the recruiter is trying to uncover. These will vary from company to company and from job to job. Nonetheless, there are certain common threads, especially in the high-tech world.

Markus Schweig of Microsoft lists the following as key questions he wants to probe when evaluating any job candidate. They reflect a sampling of some of the crucial factors that the world's great high-tech firms consider when making hiring decisions.

- Does the candidate have a strong technical background? Has he or she brought technical engineering projects to a successful conclusion before?
- Is the candidate well connected with relevant supplier companies on both a personal and business level?

- Does the candidate have a track record at two or more successful companies?
- Is the candidate used to a fast-paced business environment?
- Is the candidate able to provide appropriate structure and direction to scale new heights rapidly?
- Does the candidate know how to increase the bottom line?
- Does the candidate perform well under adverse circumstances?
- Does the candidate keep a good balance between work and personal life?
- Is the candidate a visionary leader? Is he or she a motivator?
- Can the candidate lead with politeness and patience, and get the most out of people while keeping them happy?
- Is the candidate well versed in current and future technologies?
- Is the candidate respected? Does he or she respect others?
- Does the candidate have the right sense of humor for serious business?
- Does the candidate value training for himself or herself and others?
- Does the candidate like to share knowledge, and does he or she make it easy for others to pick it up?

Wall Street Journal reporter Scott Thurm asked John Chambers, CEO of Cisco, "What do you look for when recruiting?" Chambers answered, "We want people with an ability to lead based on their life experiences, education, and values." Then he listed ten specific qualities that he and Cisco particularly value: "First, I believe, you've got to articulate to your team what you expect of them and that you're going to hold them accountable for it. . . . Second, you have to have a good team that you attract, retain, and develop at the speed that we're growing. . . . [The] third element is trust and integrity. . . . Fourth is industry knowledge. . . . Fifth is teamwork. . . . Sixth is communication skills. . . . Seventh is customer focus. . . . Eighth [is] balancing between planning and reacting. . . . Ninth is people skills. . . . Tenth is, 'Can you drive Cisco's future?'"

Asking Questions

Of course, you're not there just to answer the interviewer's questions. You're there to participate in a two-way exchange of ideas—which means getting your questions answered, too.

Actually, asking questions in a job interview serves several purposes. Among other things, your intelligent, thoughtfully planned interview questions help you to:

- Gather new information about the job and the company.
- Reinforce, clarify, and correct what you know from other sources.
- Show the interviewer your interest in the company.
- Test the topics that the interviewer may want to *avoid* discussing.
- Look for inconsistencies in the company's self-portrayal.
- Demonstrate your seriousness and eagerness to learn.
- Flatter the interviewer.
- Distinguish yourself from almost every other job candidate.

Here are some examples of the kinds of basic questions you should consider asking in any job interview.

- What are your goals in hiring someone at this time?
- What are you trying to achieve in the hiring process? What are you trying to avoid?
- What kind of individual did you have in this job previously? How well did he or she work out? Why?
- What are the most important personal and professional qualities required to fill this job?

- What spells success and what spells failure in this job?
- What are the immediate challenges, problems, issues, and opportunities that the successful candidate will face?
- Who will be making the hiring decision? When? How?

Questions like these can go a long way toward helping you get a real feeling for the style and culture of the company, making it much easier for you to tell whether or not the job opportunity is right for you. Of course, the preceding samples are just a fraction of the questions that you can ask. In fact, virtually anything the interviewer asks you can be rephrased and directed back at the interviewer. You wouldn't do this in a smart aleck or sarcastic fashion, of course. But in an interview, turnabout is really fair play. For example, an all-too-typical interview question you might hear is, "What are your strengths and weaknesses?" Why not ask the interviewer, "What are the main strengths and weaknesses of this company [or this division, or this marketing group, or this management team]?"

Smart interviewers respect candidates who are prepared with smart, relevant questions. "Ask twice as many questions as the interviewer asks," Ron Dobson of ADP Canada advises. "Ask questions that are out of the ordinary and show you are a learner. Send signals that say 'This person asks good questions and gets to the information that helps get the job done . . . questions that get to the heart of things.'"

Dress and Body Language

A *Vanity Fair* columnist wrote this about the Silicon Valley dress code: "For gents, it's easy. Drop by the Gap for khakis and don some running sneakers. For the best effect, you'll need a chambray shirt embroidered with the logo of an invitation-only conference like Agenda. . . . It's still largely the grad-student look—jeans, khakis, plaid shirts, hiking boots, or running shoes—while older guys upgrade to leather jackets and dressier shirts. . . . Ladies, your mission . . . is more complex. Hewlett Packard CEO Carly Fiorina has upped the ante on the dress code with her passion for Armani suits. But in general, it's jeans, or Capris, a twinset, maybe some Dolce & Gabbana slides for the workday. Accent with a pashmina . . . and forget the big jewelry, at least in Seattle."

But here's an important caveat. Once you have the job, jeans may be fine. But on an interview, a neat, businesslike appearance is necessary. The safest rule: Dress one notch better than the employees you see walking the company's halls, but don't let your dress be more interesting than you are. Be clean and well groomed. Nix the frayed, the soiled, and the scuffed. Straighten your hair out and trim it if necessary. Don't overdo the fragrance or the jewelry. Trust me, concessions like this won't compromise your individuality or personality, all they'll do is eliminate a possible distraction from how smart, talented, dedicated, and enthusiastic you are.

Even more important than your appearance, however, are the unspoken messages you send by your behavior and demeanor. "We still look for good eye contact and a neatly pressed shirt," says Charlie Wallace of Heidrick & Struggles, "and a degree of confidence, enthusiasm, energy and focus—without arrogance."

Here are a few specific tips you can follow that will contribute to a positive impression.

- Sit appearing relaxed. Lean back slightly; cross your legs, one knee over the other, if you find that posture comfortable. Position your arms asymmetrically (rather than rigidly identical); simply placing one arm on the armrest of your chair and the other across your lap makes you look more relaxed. And when you appear comfortable and confident, the interviewer assumes you are comfortable and confident—and you make him or her feel the same.
- If you have a choice of seats, pick the chair farthest from the door. Most people pick the one closest to the door, as if to enable a quick escape. Seize little opportunities like this to set yourself apart from the crowd.
- Try to wear a slight smile; avoid smirking, scowling, frowning, or an overly serious stare. Look the interviewer in the eye, and speak up so you can be heard.

- React to what the interviewer says. If he or she says something that's genuinely funny, don't be afraid to chuckle. If you hear something that impresses you, say so. If you're surprised by a question, comment on it, saying, "That's a good question! I don't think I've heard that one before." Not only does this response to an unexpected question imprint you on the interviewer's mind as a genuine human being, but it also buys you a few valuable seconds to think before responding.

You've done a good job on paper and an even better one in person—but there's much more before you take the plunge.

9

BEFORE YOU TAKE THE PLUNGE

"You can research, research, research before going in and still not really know what you're getting into."

—JANICE AUNE, CEO, Onvoy

Looking Past the Corporate Hype

When you've been focused on a job search for some time, it's natural to want to jump eagerly at the first good opportunity dangled in front of you. And hiring companies will do all they can to increase that desire. In order to lure the best job hunters, recruiters depict their companies in mouth-watering terms. Here's a sampling of claims from some actual recruiting ads.

> We meet today's most compelling technological challenges . . . redefine the way the world works . . . are a dynamic, team-oriented environment offering challenging projects and diverse ideas . . . strive to be a no-office-politics company . . . are a venture-capital-funded, first-class pre-IPO technology company.

> **We have an experienced and well-connected management team . . . give you a chance to run with the ball and the attention you deserve without being micro-managed . . . are located in a mecca for camping, hiking, boating, mountain climbing, fishing, windsurfing, theater, professional sports, and the arts.**
>
> **We are successful people who get the job done and like to have fun while we are doing it . . . want people who refuse to accept boundaries . . . constantly push the cutting edge . . . have a passion for making an impact . . . reject conventional thinking—and if you do, too, you should be a part of our team.**
>
> **We have enjoyed eighteen consecutive quarters of greater than 70 percent year-over-year revenue . . . give our people every other Friday off because we, like you, believe your time is an invaluable investment . . . offer 100 percent employer-paid medical, dental, vision, life and disability insurance, incentive stock options, bonus, 401(k), and a generous vacation/holiday schedule. And we leave the dress code entirely up to you.**

Does all this sound appealing? You bet! But don't believe what you read or even what you hear without testing and re-testing the underlying reality. This is your life we're talking about—isn't it worth the effort?

Even professional recruiters recognize the existence of a gap between company promises and reality. In the search world, people refer to "offered job specifications," which are differentiated from "unoffered job specifications." The former are used to entice and attract great candidates; the latter are what candidates really finds on the first day at work.

Successful recruiters are talented salespeople. I've heard headhunters boast, "If we want you, we'll get you, even if you say "No" at first. Once we get our claws in you, there's no escaping." One recruiter told me, "It's sort of like a roller-coaster ride. As soon as you sit down, they push the bar down across your lap. You can't wiggle out, and you can't get off. We're that compelling."

Even a company CEO confessed, "I can really sell the idea of working here. I know how to make this company sound like all the things employees want to hear. But I know what it's really like, and I feel a little guilty sometimes." Yes—but not guilty enough to turn off the snow job the next time a hot candidate shows up.

You've probably experienced this kind of selling job yourself. Think back: Have you ever joined a company with high hopes, only to discover, once you got inside, that it wasn't what you expected, what was promised, or what you were sold? Don't let that happen again.

Questions You Need to Ask

When you receive a job offer—or, with luck, two or three or more job offers—take the time to do your homework before making any commitment. Don't let anyone pressure you into a quick decision when it comes to your career. And don't feel guilty, embarrassed, or insecure about asking for the time you need to make the right choice. After all, in the words of Wes Raffel, a general partner with Advanced Technology Ventures, "This is about business, this isn't about friendship."

If you are as good as you think you are, the hiring company needs you every bit as much as you need them—maybe more. So be picky. Go into it only with both eyes wide open. One software company CEO advises, "A serious job candidate should spend a week analyzing the business before accepting an offer. I've seen a lot of companies who don't portray themselves honestly. . . . A twenty-five-year-old candidate could be fooled because of his or her idealism. I say, 'Buyer beware.'"

In this chapter you'll find a checklist of questions to help you make sure you've covered all the important issues that you need to consider before accepting any job offer. Ask these questions of the search consultant, the executives you'll be working with, your original referral source, and your connections in the business community. Supplement their answers with information from the company's Web site and from company press releases, coverage in the news media, and the opinions of Wall Street analysts.

Note how company spokespeople react when you raise these questions. If they seem to be bothered by your thoroughness and frankness, you may want to think twice about signing on with them—perhaps they have something to hide, or maybe they are simply accustomed to treating job hunters as powerless supplicants. In either case, a red flag should be raised. One CEO says, "If a job candidate asks too many questions, I think they might be difficult to deal with once they're on board, so I'm leery." The comment suggests a company culture that values passivity and meekness rather than forthrightness and courage.

By contrast, recruiters from any *good* organization will value the research effort you put in to get to know them. Your well-thought-out questions tell them a lot about you, all of it positive. If they're truly business-savvy, they will recognize that this is a demonstration of how you will act on their behalf once you're on board—that you are a person who questions, investigates, studies, and examines the facts before making any important decision.

The time and effort you invest in company research will not be wasted, even if you end up declining the job offer. At the very least, you'll have deepened your knowledge of the business world and perhaps gained insight into a company you might work for some time in the future. You may even stumble across a great investment opportunity in this way. One venture capitalist comments that evaluating a company you're considering joining "is like choosing one to invest in. Too many people act like the average day-trader, who goes for whatever is hot. The truth is that if it's on television or in a magazine, it's B.S." As smart investors do, take the time to dig beneath the superficial image of a company before you throw in your lot with them. Here are some questions you need to ask as you go about the task of evaluating companies you are considering working for.

ABOUT THE COMPANY'S TECHNOLOGY

This checklist provides questions that focus on the technology around which a particular high-tech company is based.

- What exactly does the company do? What products or services are provided to customers through the company's technology? How exactly does the technology benefit people?
- How thoroughly does the company control the technology, either through ownership of patents, licensing agreements, or dominance of the marketplace? How soon will rivals be able to steal market share from the company by offering alternative products or services?
- Is it, was it, or will it be the hottest technology in its field? Where is the technology in the cycle that runs from invention to widespread adoption to obsolescence?
- In what markets—geographic, demographic, corporate—is the company particularly strong? Which groups of consumers are willing and able to pay for the company's technology? How quickly are those markets growing?
- How many future applications and developments of the company's technology are possible? How big (in terms of population and potential sales) is the possible market for the technology?

Of course, it's impossible to predict the future with certainty. But your goal is to try to look ahead three, five, or ten years down the road. How great are the growth prospects for the company you're considering? The faster the company grows, the more your stock options will grow in value, and the better your chances for promotions, raises, and bonuses.

Ideally, you want to join a company that is part of an industry sector that is significantly bigger than the company itself. A $50 million company with clear access to a growing $50 billion market probably offers greater career opportunities than a $10 billion company that already dominates a static $10 billion market.

Ask people from the company to comment about the growth prospects for their technology. If the responses you get are detailed, clear, intelligent, consistent, and plausible, the company may have a real shot at being one of the big technology winners of the coming decades.

ABOUT THE COMPANY'S FUNDING

The sources and forms of the company's funding are important for many reasons. Among other things, they affect the degree of control that remains with the management team. Typically, a venture capital (VC) fund owns a percentage of the business, puts one or two partners on the board, and ends up with a lot of say in how the company is run. By contrast, if the working capital is borrowed from a bank, the founders will simply be charged interest on the loan, and they'll retain a great deal of independence. As a manager within a company, you need to consider how comfortable you might be operating with the sharp eyes of a venture capital firm second-guessing your decisions.

An even more important issue is the question of what will happen to the business (and you) if the company runs out of money. Venture capitalists withdrawing their support from funded companies helped cause the dot-com demise of 2000. As the high-tech bubble of the late 1990s gradually deflated, the demands placed on company managers steadily escalated. Commenting on the current climate for high-tech businesses, one venture capitalist says, "It's no longer enough just to show a profit. Now a business has to prove it can produce a 10 percent profit margin on a consistent basis before they will invest more money into the company."

And when venture capitalists withdraw their support, it often happens precipitously, leaving hundreds of employees high and dry with little or no advance warning. One ex-employee of a de-funded company remarks, "They didn't do it at the last minute. They did it at the last second—and I mean that literally, not figuratively."

Questions you need to ask about a company's funding include the following:

- Who owns the company? Is it privately held or publicly traded? If the former, what combination of individual investors, VC firms, institutional investors, and/or corporate partners has provided the company's start-up capital?
- If the company is largely funded by venture capital backers, is it a no-name VC firm or a well-known firm with a significant "halo effect"?

- How many rounds of financing has the company gone through? Is the company going to run out of money if it fails to obtain additional VC backing? If so, when?

ABOUT THE COMPANY'S PROFITABILITY

If you don't have a financial background, you may want to ask an adviser to explain and evaluate the company's business and accounting practices. Some high-tech firms—including some that are only marginally profitable—are based on sound business models that hold out the promise of significant profitability in the foreseeable future. Others use "creative" accounting practices to create the illusion of growth and profitability without the reality.

John Hashman, CEO of NextCard, has noted an increasing sophistication on the part of job candidates when it comes to evaluating high-tech businesses. "Two years ago," he remarks, "about the only things that prospective employees asked were, 'What's your idea? Do you have funding? And do you give pre-IPO stock options?' Now they want to know your business plan, what the analysts think of your company, and how long it will be before you're profitable." This new awareness is all to the good.

Questions you should ask about a company's profitability include the following.

- Who are the company's customers? How broad and deep is the customer base? Is a large percentage of the company's revenue derived from a small handful of customers?
- How quickly is the customer base expanding?
- How does the company make money? Does its business model make sense for the long term?
- Does the company have multiple revenue points—that is, a variety of ways of generating sales dollars? How vulnerable are those revenue points to outside competition?

- If the company is currently not profitable, what is its projected path to profitability (P2P)? How soon does management expect the company to be in the black?
- Are the profit projections based on solid numbers or on overly optimistic extrapolations?
- What is the pattern of profit and loss figures in recent years and quarters? Have profits been steadily growing in recent quarters—or, at least, have losses been steadily declining?
- If the company is publicly traded, how does the current stock price compare to the price at the time of its IPO? How does the company's valuation compare to that of closely competitive firms? How do Wall Street analysts rate the company's shares in comparison to those of competing firms?

ABOUT THE COMPANY'S CUSTOMERS

Ultimately, of course, customers are the ones who make any company profitable. A company that forgets this truth—that fails to respond effectively to the needs, desires, and tastes of its customers—will probably lose market share to more responsive competitors and in time will surely fail.

Many companies, unfortunately, lose sight of the central importance of the customer. Try to get a sense of how the employees at a company you're considering stay in touch with customer needs. Finally, if you sense that customers are treated as adversaries, or with disdain—if they seem to be spoken about as "those idiots who don't understand how to use our great products"—think twice about joining the company. Firms that look *down* on their customers usually find themselves looking *up* at the competition—and generally sooner rather than later.

Get answers to the following questions about a company's customers.

- How well do the company's current products and services meet the needs of customers? How highly are they rated by independent consumer or business information sources? How competitive are the company's prices, policies, and customer services?

- How well do the company's sales, marketing, production, design, and management personnel understand their customers' concerns? How effectively have they anticipated future changes in customer preferences and tastes?
- What are the steps involved in delivering the company's products or services to customers? What kinds of intermediaries (middlemen) play a role? What share of revenues do the intermediaries demand? How significant is the risk that one or more key channels of distribution could be closed off (for example, by being purchased by a competing company)?
- How often does the headquarters staff meet with and listen to real, live customers? How are products and services tested, evaluated, and improved both before and after their initial launch? How are customer complaints and problems handled?

ABOUT THE COMPANY'S MANAGEMENT

You have to find out whether the company's management team is made up of "A" people—leaders who are capable of exploiting even a small market niche and developing it into a major opportunity. "A" people attract "A" people, and in the process they build great companies together.

There's always a story to be uncovered concerning the management team at a company. Sometimes the founders of the company remain in charge well into the firm's second or third decade, or even longer; in other cases, founders are pushed out early in the company's history; in still others, the founders enter into a partnership with professional managers who are brought in to help run a growing organization. Finding out what has happened to the management team at a company you are considering joining, how those decisions were made, and how those changes have affected the atmosphere and culture at the firm will give you a greatly enhanced sense of what it's like to work there.

Every company's management style is unique. Some firms are relatively autocratic, hierarchical, bureaucratic, rigid, strict, and numbers-oriented. Others are democratic, open-ended, loosely organized, flexible, humanistic, even chaotic. At some companies, a small handful of top executives make most of the decisions; at others, strategy "emerges" by consensus or by the combined actions of everyone at all levels. In some companies, the top leaders are highly visible, walking the hallways, chatting with workers, lunching in the cafeteria; in others, they appear only on ceremonial occasions, reading from scripts. Some CEOs pride themselves on being charismatic, dynamic leaders who love to deliver inspirational pep talks and try to imprint their personalities on everyone at the company; others are apparently reticent and shy, exerting influence on the company indirectly and via suggestion.

It's important for you to evaluate the leaders at a company you're considering joining. Of course, scrutinize their management backgrounds—executives who have succeeded at several other companies probably have a better chance of leading their current companies effectively. But also consider their personal leadership styles and how well these mesh with your personality. Based on your own past work experiences, you may be able to anticipate the kind of leader you're most comfortable working with. Look for this type of executive at any company you're evaluating.

Consider the following questions about the people who run a company.

- Who founded the company? What kind of business and educational background did the founders come from?
- Who are the CEO and the other members of the current executive team? What are their business and educational backgrounds? What motivates and interests them? Is there strong camaraderie among the key people? What is the turnover rate among the members of management?
- Who is on the board of directors? How actively is the board involved in corporate decision-making?
- How does management describe the company's mission, philosophy, and values?

- What decision-making and leadership style is employed by management? Is there any apparent disconnect between the *avowed* management style and the *actual* management style?
- How does the company attract and retain key employees? How do financial rewards rank on the scale of key motivating factors? How does this compare to such less tangible rewards as creative freedom, an idealistic company mission, service to the community, a humane working environment, and so on?

ABOUT GEOGRAPHY

High-tech companies are everywhere. Although Silicon Valley was the launching pad for the boom of the 1990s, the high-tech sector has now spread throughout the country—and the world.

Wherever you are currently living, it's likely you can get into a high-tech business without going far. For some people, staying put is important; they may not want to uproot their families, or they may value the quality of life in their present location. Others find it exciting to move, or are at least willing to move if the career opportunity makes a move worthwhile.

Here are some of the questions to ask if you're considering a job that will require a geographic shift.

- What is the economic situation in the new community to which you may be moving? How high is the cost of living, especially for variables such as housing and health care? What are the demographics of the local work force, including levels of education and average unemployment rates?
- Who are the major private-sector employers in the new community? How healthy are these companies financially? What kind of reputation do they enjoy as employers?
- How do key lifestyle variables stack up in the new community? What are local crime rates like? How is the weather? What opportunities exist for participation in the arts, sports, and civic, religious, and cultural activities?

- How wired is the new community? How easy is to get high-speed access to the Internet? How much does it cost?
- What will traveling to work be like? Is traffic a nightmare? Is mass transit available? How flexible are work hours? Is telecommuting (working from home) a viable option?
- How will your family thrive in the new community? What career opportunities exist for your spouse or other loved ones? What is the reputation of the local public or private schools?
- What is the ethnic, religious, and class makeup of the population? How diverse is the community? Will you and your family "fit in" comfortably with your new neighbors?

Geography definitely affects the compensation you'll require and should expect. For example, according to a July 2000 study by ERI Economic Research, the annual salary of a Webmaster with three years' experience might vary as 20 percent based solely on job location, as follows.

San Jose, CA	$64,503
Los Angeles, CA	$61,138
San Diego, CA	$57,407
Portland, OR	$56,761
Denver, CO	$57,632
Austin, TX	$52,238
Phoenix, AZ	$54,036
Orlando, FL	$52,780
Boston, MA	$60,437

Of course, the cost of living in these diverse areas determines the true value of these disparate salaries.

For many in high tech, northern California's Silicon Valley remains the mecca. The wealth of great companies and creative minds located there offsets most of the disadvantages: traffic congestion, high taxes, poor schools, and high living costs. In the Silicon Valley area, a one-bedroom, one-bath apartment carries an average rent of $1,645 per month, while the price of a "starter" home can average $500,000. In search of affordable housing, Silicon Valley workers have scattered all over the state's sprawling communities, and some commuters drive a hundred miles each way.

But, as one local says, "There is charm in the fact that you can be totally grubby, make tons of money, drive a great car, and be in on the latest technology." If that's your idea of heaven, then by all means head for Silicon Valley.

ABOUT THE COMPANY'S CULTURE

The most important job of management is to get and keep the organization in the black, so that employees have some place to go to work. The *second* most important job of management is to make the organization a decent, enjoyable, productive, and creative place to work—in other words, to foster and nurture a positive corporate culture.

If your most important work values aren't shared by a company you're considering, think twice before signing on. This issue is so important that you shouldn't rely on the accuracy of what you're told by the hiring executive or recruiter. You need to speak to your own business contacts, present and past employees, and company vendors and customers. Compare the company's self-image with its outside reputation. (The latter is often more accurate.) Both you and the company benefit if the cultural fit works.

- How do the company's leaders describe the company's culture (in recruiting materials or the annual report, for example)? How does this compare with the way rank-and-file employees, former employees, competitors, customers, and suppliers describe the culture? (A major difference here may forecast trouble.)
- Are employees treated like partners, with respect for their individuality, creativity, and personal needs? Or are they treated like interchangeable parts, "troublemakers," or wayward children?

- What is the working environment like? What kinds of working spaces do most employees occupy? How great a gap is there between the accommodations of the top executives and those of the lower-level employees? How well are shared spaces (meeting rooms, lounges, cafeteria) maintained and supplied?
- What is the mood of the offices like? Does a visitor notice joking, laughing, music, conversation? Or is the atmosphere tense and hostile?
- How do the employees dress? How do they decorate their offices, desks, cubicles, and other working areas? Is there an atmosphere of personal expression or one of regimentation and corporate control?
- How does the company help employees develop professionally? What investments are made in training and education? How are mistakes viewed?
- How do employees at various levels describe their work and the company's mission? Do most employees regard their work as "just a job"? Do they view themselves as "changing the world"? Or is the prevailing attitude something in between or altogether different?

Today more and more companies—especially in the high-tech sector—recognize the importance of corporate culture for attracting and retaining top-notch employees, and they are deliberately tailoring their company policies and practices in a quest for greater creativity, openness, and fun. In fact, when Andrew Raskin, CEO of Gazooba Corporation, recently searched the online jobs database at Monster.com using the keyword "fun," he got 748 job listings in San Francisco and another 444 in New York City. These are companies that—like Gazooba—are purposefully using the pursuit of fun as a recruiting tool.

Here are some examples of the varying corporate culture strategies being pursued by high-tech firms.

- When prospective employees walk into the headquarters of EarthWeb, they are shown the meditation room, the nap room, the piano, and the yoga room. The cubicles are twice as large as in most other companies, with dividing walls low enough to let every employee enjoy the sunlight and the view.

- The editors at *Working Mother* magazine selected Pearson Education as one of the best companies to work for based on their flexible work arrangements, opportunities for women to advance, generous parental leave, child and elder care referral, and adoption assistance. "Since many of these programs have become a part of our culture over the last several years, it's easy to take them for granted," says Gloree Parker-Roden, senior vice president for enterprise services at Pearson Technology Centre. But employees at thousands of other companies can only fantasize about such benefits.
- At Microsoft, almost every employee has an office, not a cubicle, and walls are adorned with museum-quality art. One Microsoft employee told *Fortune* magazine, "Microsoft is a country. I moved here, and it is home to me."

What matters most to *you* as an employee? Is it attractive and spacious working quarters? Generous benefits? "Green" environmental policies? Flexible scheduling? Absence of bureaucracy? A commitment to staff education? Make a list, then investigate the companies you're considering joining. Make sure that the business family you become part of is one you'll be happy and proud to call your own.

10

NEGOTIATING YOUR DEAL

"My father used to say, 'If you don't do nuthin', nuthin' happens.'"

—MICKEY GART, president,
Gart Development Group

Use Your Power

When negotiating for your next job, you have more power than you think you do. This is especially true at the moment when a company offers you a job. They've chosen you from among dozens or even hundreds of candidates—they decided (or convinced themselves) that you are the best choice, the person who can help the company (or the division, or the department) scale new heights of achievement and profitability. At that moment, there's a halo of great expectations around you. If the recruiter fails to rope you into a contract, it will be perceived as a disappointment, a failure. Thus, it's likely that the company will be willing to do almost anything reasonable to close the deal.

At the same time, you have to be savvy in handling the negotiation process, whether negotiating with a recruiter or a corporate hiring executive.

Defining what's reasonable takes sensitivity and intelligence; so does maintaining the aura of good feeling between you and the hiring company even as you arm wrestle over terms. You have power, but you must use that power with diplomacy and delicacy if you want to retain it. However, make no mistake: Compensation negotiations are expected. In a poll of a thousand human resources professionals, the Society for Human Resources Managers found the following:

- Ninety-two percent said salaries are generally negotiable.
- Eighty-two percent admitted that the first salary offer they make is just a starting point.
- Seventy percent said they are comfortable negotiating salary.
- Only 21 percent of job candidates are equally comfortable negotiating salary.

The purpose of this chapter is to help you increase your comfort level when negotiating on your own behalf.

Even though recruiters expect to negotiate salary, that doesn't mean they will make it easy on you. Every company is always looking for ways to cut costs. The volatility of the stock market, the pressure to increase profits, overseas market turmoil, and increased competition at home are all reasons (or excuses) for companies to say no to your salary demands. And they are a lot more practiced at the negotiation process than you are.

If an outside recruiter is involved, the process may go more smoothly than when you must negotiate directly with the hiring firm. After all, the recruiter contributes a degree of objectivity; his or her pride and ego are less invested than is the case with either of the principal parties. And with a fee on the line, the recruiter has a strong incentive to make the deal happen. "Negotiating is confrontational by nature," observes Dave Hardie, a partner with Heidrick & Struggles. "I don't want candidates and clients disagreeing and arguing even before they start working together. I'm the middleman. I hear when either party is upset or frustrated, and I can usually take care of it."

"Money is an explosive issue," adds Fred Taraschi, a managing director with Ken Clark International. "I try to fairly play both ends and meet in the middle. What I want is smiles and handshakes on the first day of work. I don't want the company to be thinking, 'You better be good because you asked for so much.'"

Naturally, you don't want that either. Your objective in negotiation should be twofold: to achieve a favorable set of contract terms for yourself, and to create a sense of victory and satisfaction of both sides. Your goal should be to "win" a negotiation, but not at the expense of fairness.

What Are You Worth?

Information is power, especially when it comes to negotiating. There's no better tool for getting a better deal than having access to solid facts that help you determine what you're really worth on the open market. Here are some tips for making information into one of your negotiating strengths.

Make a habit of knowing how comparable professionals are paid. Take time periodically to research the pay of others who are contributing similarly to their companies so as to see how your pay compares. You should know your current company's pay range for your position as well as industry standards in your area. There are plenty of places to research salaries, including Internet sites such as www.salary.com, professional associations, executive recruiters, venture capitalists, as well as your friends, mentors, and other personal sources.

Some human resources departments order their employees not to discuss salaries; some even make it a ground for dismissal. Actually, such threats are not legal. Federal labor laws dating back to 1935 prohibit employers from forbidding discussions of pay. Handle conversations about salary with tact and discretion, by all means—but don't be intimidated into believing that they are unethical or improper. They're not.

Unfortunately, many high-tech workers don't really know what kind of salary they deserve, which results in unfairness. "There is a need for people to understand what they are worth for the work they are doing," remarks Andy Schlossberg, lead animator with romp.com. "For the online worker, there is a serious lack of pay standards. A great many people are being overpaid for poor work, while some amazing talents are being underpaid."

Of course, the kind of information that's publicly available—average salary data for specified positions, for example—will carry you only so far. Your pay potential is also tied to your past and future contributions to the organization, the specific job description you'll be filling, the level of demand for people to fill your position, the number of people with the skills needed to fill it, the number of candidates in a specific geographic area, the strength of the overall economy, and the financial status of the company. You should weigh all of these factors in determining the salary to ask for.

As a rule of thumb, many executive recruiters today say that your salary should equal three times your age. Thus, if you're a forty-year-old middle manager making $100,000, you're behind, according to their calculations. Does this mean you're underpaid? Maybe. It could also mean that your skills and productivity lag behind those of your peers, so that you don't deserve peer-level compensation. Deciding what's fair and unfair can be difficult; it requires a high degree of realism and self-awareness as well as grounding in solid fact.

The good side of doing your research and knowing what other people are paid is that, when analyzed objectively, such data can help you determine your comparative worth. The bad side is that you can go a little crazy worrying about inequities. It's true—in the real world, salaries are sometimes arbitrary, lopsided, and unfair. If the inequities are in your favor, that seems okay. It doesn't seem okay if they aren't!

Here are typical salary ranges within a single company for people hired at different times but doing basically the same work.

- Project manager: $65,000 to $107,000
- Strategist: $78,000 to $113,500
- Vice president: $105,00 to $177,500

Is there a just and logical rationale for each of these salary variations? Probably not. How seriously you take the issue is up to you.

One inequity that can seem particularly galling is when companies pay new hires more than their veterans. Sometimes the discrepancy is as much as 30 percent. Companies rationalize that it's necessary for them to pay a premium to get someone to leave another organization. Another justification is that the new hires come in with an "increased demand for performance," as one HR person put it. Nonetheless, when loyal employees find out about the new-hire differential, it creates resentment, undermines morale, and diminishes team spirit.

One result of the new-hire differential is that many working people feel they *must* change companies from time to time in order to keep up with their industry peers. What happens is sometimes deeply ironic. It's not unheard of for an employee to become dissatisfied with his salary at Company A, shift to Company B in exchange for a promotion and a 30 percent raise, and ultimately return to Company A for yet another promotion and an *additional* 30 percent raise.

It should only happen to you.

Beyond the Bottom Line

Remember that "compensation" doesn't only include salary. You must also consider the monetary value of any high-level executive benefits you may be offered. Some of the perks that high-ranked managers have received include the following.

- Access to an on-site health club
- The services of an executive dining room or even a private chef
- A weekly massage
- Use of an on-site meditation room or music room
- Monthly house cleaning
- Concierge services for laundry, shopping, and other personal chores

- Company-paid trips to warm climates
- A laptop computer for home use
- Discounts at popular retailers, such as J.Crew or Ann Taylor

During the high-flying dot-com years, some companies vied to out-perk one another. Although those days are gone (for now), top executives who are in demand will always be able to ask for out-of-the-ordinary treatment. "Technology companies like to imply they originated these benefits," sniffs a director at a Fortune 500 firm, as she pauses to hand her shoes to the free in-house shoeshine concierge. "But we've had an on-site salon, in-house dry cleaning, discount tickets to sporting events, and car detailing for the convenience of our employees for years."

If you're not yet in the executive stratosphere, remember that even basic benefits are worth a lot. The overall compensation for most professionals includes significant noncash value. For an employee earning $100,000 in salary, basic benefits typically add up like this:

BENEFIT	**DOLLAR VALUE**
Health insurance	$3,034
Dental	$120
Life insurance	$300
Disability insurance	$1,200
401(k) plan	$3,000
Social Security/Medicare	$5,739
Two weeks paid vacation	$3,848
Three paid personal days	$1,156
Worker's compensation	$2,000
Unemployment compensation	$3,136
Total	$23,533

In addition to such monetary benefits, many people consider such intangible rewards as respect, appreciation, recognition, creative freedom, autonomy, and flexibility to be as important as salary and benefits. It's up to you to let your manager know what motivates you so your company can reward you accordingly.

Above all, expect to be rewarded in accordance with the contribution you make to the company. As executive recruiter Alan Seiler says, "Feed the eagles, starve the turkeys—that's the whole idea of a reward-based system, where the people who really deserve to be taken care of are taken care of." If you really want to be at the top of the compensation heap, do all you can to be an eagle rather than a turkey—and then make certain you're recognized as one.

Negotiating Tactics and Strategies

Let's assume you've developed a fairly clear idea of what you're worth on the open market. How do you use this information when the time comes to negotiate your next deal?

You may have a specific salary figure in mind, but you don't necessarily want to put it on the table first. You may be asking less than they expect to pay, in which case you may end up leaving money on the table if you volunteer a figure first. Instead, ask the recruiter the pay range they have in mind for the position. Then look for skills not included in the position description that you bring to the company. These can serve as a justification for you to get paid more.

If you must state a dollar figure first, give a very high one, never the minimum amount. There are basically just three possible responses that the hiring firm can offer. They may say, "That's way out of the ballpark"; they may say, "We'll see what we can do"; or they may say, "Okay." Obviously, the third answer is a clear win for you. The second answer is satisfactory, too; it opens the discussion on terms that are favorable to you. Even if they give you a flat-out no, it needn't damage the tone of the discussion. You can always say, with a smile, "I'll take that as a maybe." Adopting an attitude of good cheer about it all lightens the heaviness of the situation for you and them. At the same time, your answer lets them know that you aren't dropping the issue. Here are some other helpful salary negotiation strategies.

- **Be pleasantly aggressive.** While maintaining a positive, friendly attitude, remember that you owe it to yourself to get the best package possible. This applies starting with your very first job. After all, since most companies factor your current salary heavily into their planning for current salary negotiations, every deal you negotiate establishes a floor for every deal that follows. So imagine the effect of being pleasantly aggressive about salary starting at age twenty. By the time you're forty, you'll be far ahead of your peers, some of who may only be developing the gumption to speak up by then.

- **Sleep on it.** Whenever you receive a compensation offer, slow down, take a deep breath, and ask for time to think about it. Sleeping on the offer before you accept, decline, or reopen negotiations will benefit you in several ways; it gives you the opportunity to think about the deal and consider each aspect of the package in relation to the whole; it gives you a chance to think of new and creative ways of bridging whatever gap may exist between you and your prospective employer; and it permits you to discuss the offer with your spouse, best friend, or career mentor. In addition, acting a little hard to get rather than overly eager has a way of increasing your value in the company's eyes, thereby enhancing whatever leverage you already enjoy. Don't let anyone pressure you to decide immediately, but set a reasonable deadline for responding. Twenty-four hours is minimal; forty-eight to seventy-two hours is fairly common.

- **Negotiate professionally.** Present yourself in a manner that supports your case. Be likable, so they'll be willing to talk to you about this again. Your friendly manner and reasonable approach can help get you bumped up over the predetermined salary range. If they like you and therefore want to make you happy, they may decide to bend their rules. If they don't like you very much, they won't be so concerned about making you happy.

Decide in advance the rock-bottom figure you'll accept. It must be a number below which you'll genuinely be willing to walk away. If your determination is shaky and you end up accepting measurably less than your minimum figure, you'll probably end up regretting your decision and resenting your new employer for lowballing you.

Above all, try to remain objective rather than emotional throughout the negotiation process. "Don't take it personally. And don't take it home," advises Christine Carey, vice president for Web services at AMI Visions. And keep repeating this mantra at *every* negotiating session: "I know you'll be fair with me." They just might.

- **Get it in writing.** "It ain't over till the fat lady sings, even in contract negotiations, 'cause anything can go south," warns Nancy Albertini, CEO of Taylor-Winfield. "Even a signed deal can come unraveled." It's true that, in strict legal terms, an oral agreement is every bit as binding as a written one. But not everyone in business practices an honor-the-contract rule.

Doing business on a handshake is seldom possible today. As one young executive advises, "Get *everything* in writing. This is especially true when it comes to job offers. It doesn't matter whether the offer comes from a close personal friend or a complete stranger. You never know when someone will try to squeeze you out of money you're owed."

The documentation of your job offer may take any of several forms. In some cases, a formal contract will be drawn up, in which case you'll want to consult an attorney for advice and guidance. In other cases, a less formal "offer letter" or "term sheet" will be presented. Note that such a document is usually considered a legally binding contract even if it isn't labeled as such, particularly if both parties sign it. So assume that any document you receive and agree to will be legally binding. This applies even to a company handbook or sheet of boilerplate employment policies.

When you receive a written confirmation of a job offer, double-check what's in writing versus the oral understanding in regard to:

- Your salary, bonus, stock options, and other compensation
- Starting date and job title

- Stated period of time for which the agreement is in effect
- Salary review period
- Moving/relocation provisions
- Change of control arrangement (i.e., what happens to you if the company is sold)
- Vacation, holidays, sick days, personal days
- Medical, dental, life, disability, and other insurance programs
- Termination procedures and severance package
- Any other benefits or perks you've discussed

Really listen to what people say, offer, or commit to. Then make sure you see it in writing. Whenever you think, "Oh, it doesn't matter if they left that point out—we won't have a problem with it," the chances are good that you will.

The appendix to this book contains a sample boilerplate employment contract to give you an idea of what to expect from such a document.

- **Understand your post-employment restrictions.** Nowadays, more and more companies are eager to restrict former employees from activities that they view as competitive. In pursuit of this goal, they ask new employees to sign *noncompete agreements* at the time of hiring. They justify this stance by saying, "We don't want former employees to steal our customers, hire away our best workers, or start new businesses based on secrets they learned while working for us." Fine, except that some noncompete agreements go so far as to limit workers' rights to make a living once they are fired or quit.

Traditional noncompetes were geographically limited; for example, employees may have been restricted from conducting the same kind of business within twenty-five miles, say, of their former employer. Noncompetes were also limited in time—say, for three years. These traditional provisions are changing. With the global reach of the Internet, geographic limitations have become less meaningful, so that noncompete agreements are now being written to cover the whole world. On the other hand, because current technology changes so quickly, traditional noncompete terms of two, three, or ten years are now often limited to just six months.

If you're asked to sign a noncompete agreement, make sure you know exactly what it covers. There are many variations. For example, a so-called *nonsolicitation agreement* typically states that a former employee may not "solicit" any company customer, supplier, or employee for any business purpose. Thus, if you quit Company A to start your own firm, you'd be legally prohibited from offering a job to a colleague at Company A, or from trying to sell your products to a customer of Company A.

Also look into the impact of local laws on any noncompete agreement you are asked to sign. Such restrictions are viewed quite differently in various parts of the country. Where Florida courts are more likely to interpret such contracts in ways designed to protect the employer, California judges are more likely to protect the employee.

- **Be discreet.** When your deal is settled, keep quiet about it and about the dealings that led up to the settlement. Any bragging will create resentment and office politics problems you don't need. (This doesn't contradict the advice about the right of employees to discuss salary among themselves. Sharing information to be helpful is one thing; bragging is quite another. Do the first, not the second.)

If you're involved in super-big-deal situations, some companies may ask you to sign a nondisclosure agreement, which binds you to secrecy as you negotiate. Sometimes discussions are so secret that they are conducted in "stealth mode," using code names for all the principle participants. This may not really be necessary, but it does underscore the importance of confidentiality.

Leveraging a Better Deal from Your Current Employer

Sometimes it's possible to use a job offer from another firm as a way of leveraging a better deal from your current employer. Or you may want to approach your current boss for a raise even before having another offer in hand, simply on the basis of what your research into the job market has revealed. "Negotiating [a better employment deal] isn't an 'on occasion' event," says Dan Purdy of Agilent Technologies. "It's an ongoing process. At its heart is consistent hard work and constant communication with those who have the authority to reward you."

Although no company is eager to increase its payroll costs needlessly, good bosses appreciate it when employees are frank with them about any job dissatisfaction they may be feeling. "I want to know when key staff members are looking around, and I encourage them to meet with me before they go too far in the search process," says Ted Wright, CEO of bigpond.com. "It could be a salary increase issue, but it could be something else, such as training needs, promotional opportunities, or lack of recognition. I think openness and candor are critical." Sasa Radulovic, CEO of TrueSpectra, echoes the sentiment: "It's our job to watch where the compensation market is and adapt to it." Unless you talk with your boss about your needs, how will he or she know what those needs are?

To maximize your chances of getting the deal you deserve even without changing companies, follow these guidelines.

• **Pick the right time.** As in so many things, timing is important. Be careful about when you initiate compensation conversations. A potentially good time is when:

- You've just received great public kudos for your work.
- Headhunters are pursuing you with job offers.
- The company is doing well financially and getting a lot of positive press.
- The labor market in your specialty is tight, and your department is understaffed.
- Your boss is in a good mood because of recent successes.
- Another division of the company has asked about your availability.
- The budget for next year is about to be set.

Don't bother asking for a raise if your work is mediocre, even if your peers are making more than you. Don't do it if the company is experiencing tough times (that is, cutting jobs, trimming budgets, suffering declines in the stock price).

- **Plan ahead.** A high-tech executive—John—told me about a personal development plan that he and his boss agreed on before John started his job. It listed the tasks John was expected to accomplish and the rewards he would receive once those tasks were completed. "If they forget," John says, "I have my documentation binder to speak for itself." John continues, "Since I know it takes several months to get the paperwork through my manager's manager, I will refer to my binder two months prior to the typical semiannual review and drop subtle reminders to my boss regarding my performance."

John's strategy is a good example of the power of pre-planning your own compensation upgrades. Keep an open-door policy about the subject, even if your boss doesn't. Be honest and open about your salary goals and let the company know about them up front. Unless you communicate your value, remain aware of the options open to you, and pursue your goals pleasantly and persistently, you'll ultimately be left behind.

- **Be able to prove your contribution to the company.** When you're ready to ask for a raise, have clear evidence to demonstrate how you've helped increase your company's profitability, and develop your strategy around that evidence. Have figures and specific examples to support your claims. "Give me facts," says David Selden, CEO of ProBate Software. "How you compare to peers. How you've performed the last two years. Then come to a reasonable conclusion." Again, objectivity is the key, *not* emotional appeals.

Don't expect your boss to recognize your greatness. Whether he or she has five or five hundred reports, your boss may not be aware of all your specific contributions. So build your own case and document what you've done above and beyond the basic job requirements. Write summaries of courses you've taken, new skills you've developed, praise you've garnered from team members, managers, and customers, and especially the things you've done that increased revenues, trimmed costs, and boosted profits. List all the times you

- Played a major part in acquiring a significant new piece of business
- Did something to improve product quality or customer service
- Initiated or advanced the development of a new product or service
- Saved the company a significant amount of money
- Took on more responsibility, authority, or risk
- Increased the visibility of the company, its products, or its services
- Received an outstanding performance appraisal

Having done good, results-oriented work in the past is one thing; bringing more of it to the table *quickly* in the future is another. Both are important. So try to anticipate and meet the goals set by your boss or by the company as a whole. Knowledge of the corporate objectives shows that you care about the organization and want to satisfy their interests, not just your own.

Sasa Radulovic of TrueSpectra articulates this well: "If you want a raise, you can come in and say, 'I've done good work, I want a raise.' Okay. But you [also] have to quantitatively articulate the increased expectations that need to be met for that raise. A company needs to get more out of a person than the company gives."

The bottom line is simple: They only pay you more if they are getting more from you.

- **Think flexibly.** If the company won't or can't increase your salary to the level you'd hoped for, ask for something else. It's often easier to negotiate noncash rather than cash-based benefits. Even when there's a company-wide salary increase cap of, say, 3 percent, other benefits can be used to make up the difference for deserving employees. For example, you could request:

- The freedom to work at home one or more days per week
- First shot at a new project or opportunity
- Training and professional coaching
- Your next salary review on an accelerated schedule
- A guaranteed bonus or pay increase contingent on meeting specified sales or profit targets
- A match of salary to that of any new hire at the same level
- Relocation to an office you prefer
- An extra week of vacation
- A profit-sharing agreement
- Company-paid retirement planning or investment counseling
- Being the first to be upgraded to a new computer system or software package
- A plum assignment at an upcoming trade fair in Las Vegas, Hawaii, or Europe
- Face-to-face time with the CEO
- A prestigious corporate internship
- A paid sabbatical

Some people even request the right to devise their own, more creative job titles. Maybe it would make you happier to carry business cards with a cool-sounding title, such as *Chief Morale Officer, Consultant Monitor, Vice President for People, Chief Catalyst, Dreamer Extraordinaire, Talent Officer, Chief Listener, Product Evangelist,* or *Corporate Seer and Prophet.* Depending on your company's culture, if such a creative title appeals to you, it may be yours for the asking.

11

GOING FOR THE GOLD

"Sometimes they say, 'It's not about the money.' When they say that, I say, 'Bull.'"

—Executive recruiter

Money may or may not be your main career motivation, but for nearly everyone it is a factor. How much you consider enough will depend on many factors, especially where you are in your career progression.

One high-tech entrepreneur told me, "I used to dream of having a million dollars in cash. Once I had it, I decided that three million would be enough. Then four, maybe five would do it. Now I think fifteen million is all I really need."

Another woman told me that $25 million in cash was always her goal. Then she made it to $250 million. Now she is quick to point out, "There are only 1,900 women in the world with greater net worth than me." Has she stopped trying to accumulate more? I don't think so!

If money is very important to you, the high-tech arena is a good choice. In technology, if you can dream it, you can achieve it. Talent, energy, and mobility count more than seniority in creating real opportunities. In this chapter, we'll consider the best ways to make truly big money in technology.

Have Today's Hot Skill

To maximize your earning potential, be in a field where the demand is greater than the supply.

As you know, information technology (IT) jobs pay better than most other jobs because demand for talented IT workers is so great. Within IT, the relationship between supply and demand is the same as in other industries. If, for example, today's hot skill is Web development, and there are a limited number of people who are highly gifted in that field, salaries for Web developers will balloon, and no one in the marketplace will ever remain out of work for long.

Of course, like all things in technology, today's hot skill will change. As one retired thirty-two-year-old former CEO said to me, "The hot thing for two years down the road hasn't even been thought of yet."

Know which technology is up and coming and be willing to embrace that technology, but work on being a generalist in business as well. Understand how economics work, what drives profit, how consumers think, and what motivates human behavior. To make money even in a technical environment, you have to be able to communicate like a normal person while offering a special, unique, and valuable specialty.

Mark Miller, a group executive vice president with Right Management Consultants, says it well: "My advice to technology people is to broaden their perspective so they can effectively communicate with executives in the nontechnology portions of their company. Future demand will undoubtedly be very high for individuals with a technology background who can manage successfully in a general business environment."

Get Equity

Equity—that is, a share in ownership of the company you work for—is the best path to significant wealth in the high-tech sector. For most employees, equity comes in the form of stock options.

At one time, options were reserved for company board members and senior executives. Companies started the trend of offering options to lower-level employees in the early 1990s, at a time when cash-starved Internet companies needed some way of luring talent from big, established organizations. The practice ultimately cascaded through the business world, where the demand for talent in all organizations resulted in increasing demands for some sort of ownership. Today, says Jonathan Ehrlich, cofounder of Accompany, "Equity is like toilet paper; I don't think start-ups have a choice anymore."

Giving stock options to key employees is good for a company for several reasons: It attracts talent, it aligns employee interests with shareholder interests, and it helps retain talented employees who are waiting to be vested. And for the employee, the opportunity to share ownership in the firm can be a door to really significant money. But don't assume that stock options bring with them a guarantee of wealth. That's simply not the case.

A FEW KEY TERMS

Stock options differ from company to company. Employees at different levels get treated differently, although there are some basic rules that apply across the board. To help you evaluate the stock-option deals you may be offered, here are some definitions so you can "speak the language" and formulate the specific questions you need to be asking.

- **Option:** The right, but not the obligation, to buy a set amount of stock at a specific price during a specified time period. If the market price of the stock increases during the time you're eligible to exercise your option, you will profit.

- **Strike price:** The preset price that the option holder pays for the stock when the option is exercised. The strike price is almost always the market or appraised price of the stock at the time when the option was granted.
- **Exercising the option:** The act of buying the stock at the strike price. You can exercise an option by (1) paying cash for your shares, (2) swapping stock you already own for new shares, or (3) borrowing money from a stockbroker while at the same time selling enough shares to cover your costs.
- **Vesting:** The right of the option holder to exercise the option, which becomes available over time according to a predetermined schedule.
- **In the money:** When the option strike price is below the current market price of the stock, representing a potential gain for the option holder.
- **Out of the money:** When the option strike price is above the current market price of the stock, representing a potential loss for the option holder.
- **Incentive stock option (ISO):** A type of stock option on which the option holder pays no taxes until the stock is sold. After exercising the option, if the option holder holds onto the stock for a certain time period, the gain on the sale is taxed at the lower capital gains tax rate.
- **Nonqualified stock option *(nonqual)*:** A type of stock option on which the option holder is taxed on the gain at the normal income tax rate at the time the options are exercised.

As you can see, the amount of profit—if any—that you'll derive from your stock options is dependent on what happens to the company's stock price. That's one reason companies like to compensate employees using options: Workers who hold options want the stock price to rise, which may mean they are willing to work harder to make the company successful.

Of course, if the company stock tanks—even if that happens because of a general economic downturn or an irrational swoon in the stock market—the option-holding employee suffers. As Temma Enrenfeld writes in *Newsweek,* "Here's a lesson in lower math. When a stock drops from $250 to $30, you lose 88 percent of your money. To get back to $250, you stock now has to gain 733 percent. And that's just to break even. How many of your knocked-down loves do you think have such fire in their heels?"

Thus, in order to determine the potential value of a stock option, you really have to take out paper and pencil and do some calculating. And if you're being offered a job in which a large portion of the compensation comes in the form of options, you need to ask yourself: Will I still make enough money if the stock goes nowhere?

ESSENTIAL QUESTIONS

Here are some of the clincher questions about stock options that many people have kicked themselves for not asking.

- *When will my options vest?* The shorter the time period, the better for you. Sometimes, to get the hot talent, companies agree that the options will vest immediately. In other cases, the time period is quite protracted. One CEO had an option for 3.9 million shares that would be vested over a period of twenty years. No wonder he moved to a company that gave him an option for 3 million shares that vested over just four years.
- *After the company goes public, when will restrictions on insider stock sales expire?* In other words, when will you be free to sell your shares and thereby turn your potential paper wealth into cash?
- *If the company is sold, how will options be treated? Will they vest immediately?* Without this provision, you could end up with nothing if the company is sold and you are laid off or choose to quit because you don't like the new owners.
- *Will options be repriced if the share price falls?* The usual answer is no, but there have been exceptions—and if you can get this concession, it will be potentially a huge benefit for you.

- *What number of shares currently exists?* Divide the putative value of the business by the number of shares. You must try to put a realistic value on the whole company so you can compare the stock against that of comparable companies.
- *What are the tax consequences of exercising your options?* Review the details with your tax accountant.
- *Who gets to participate in options at the company?* Management decides how many options to give to employees and who will receive them. Workers at different levels get different amounts, with other factors, such as length of service, also playing a role.
- *What are the company's real prospects?* You need to know what your options could be worth given different exit strategies. A single California law firm is currently pursuing fifty fraudulent-inducement and breach-of-contract suits due to broken stock-option promises.

Striking It Rich with Stock Options

In general, only founders and senior executives who join a firm early in its history get enough share options to strike it rich. A typical stock option breakdown for a company with a $1 billion valuation might look like this:

EMPLOYEE	SHARE	STOCK VALUE
CEO	8.00 percent	$80,000,000
VP	1.60 percent	$16,000,000
Director	0.32 percent	$3,200,000
Manager	0.06 percent	$640,000
Entry-level professional	0.01 percent	$128,000

If you're determined to build your fortune via the stock-option route, you need to select the technology company you'll work for carefully and then get in early. Most start-ups go through the following three financial phases.

1. Pre venture capital
2. Post venture capital (or pre IPO)
3. Post IPO

Generally speaking, while you may receive more options if you are hired before a company attracts venture capital funding, your options will be worth more the closer the company is to going public.

John Borden, COO of LearningSoft, has these observations about the stock-option game: "Many tech executives find the best risk/reward balance at the VC investment stage, and in fact, they are often recruited by the VC firm. A C-level executive (that is, CFO, COO, CPO, CTO, and so on) at this stage can earn a decent salary ($150,000 to $250,000, or more) and receive considerable equity compensation. The big win here, of course, is the stock options granted in a pre-IPO company at a strike price commensurate with what the VC firm paid. In some cases, the amount of equity granted to a highly sought after manager can exceed that issued to the founding team. Upon a successful exit, this can mean millions for the successful executive."

The key is to think like an investor going in. After all, you are investing your most precious commodity—your time. Rather than relying mainly on stock options for your compensation, consider this alternative, which has been seen more frequently since the stock decline of 2000: Ask for a higher salary and invest in company stock on your own.

The late-1990s dot-com boom produced enormous paper wealth for thousands of high-tech workers loaded with stock options. Today, much of that wealth has vanished. One company in Silicon Valley that had 100 millionaire employees (out of 330 total employees) has only 20 today, because of the company's depressed stock price.

The Wall Street Journal recently published a front-page story about the "90 percent club," dot-com millionaires whose paper fortunes had plummeted 90 percent or more from the market highs. People who took equity in companies such as eToys, Webvan Group, Ask Jeeves Inc., and Priceline Inc., were down on average a staggering 95.7 percent.

No wonder Jack Falvey, CEO of MakingTheNumbers.com, remarks, "Stock options are gambling chips."

Does this mean that the equity path to wealth is no longer passable? Not at all. It simply means that the old boomtown days of *instant* wealth are gone—probably forever. Instead, expect to build your high-tech fortune more gradually, with a fair amount of hard work (and a little luck) as part of the equation.

Many younger people in the San Francisco Bay area use the following career strategy: They work for a promising high-tech company for a year or two, until their stock options are 25 percent to 50 percent vested, then they move to another good company for two years, then another, and so on. After a decade or so, they have a diversified portfolio of stock from several companies, any one of which might be worth a lot one day, rather than having all their net worth tied up in a single stock. Of course, this strategy also gives them experience with different companies, which can be an asset in itself.

It's a worthy strategy to consider for anyone who is young, talented, and willing to take on risk.

Ownership Opportunities at Private Companies

So far, we've looked at equity in terms of stock options—that is, the opportunity to own a share of a publicly traded company. But if a company is privately held, there may still be options for you to share in its ownership, literally or figuratively.

At privately held ProBATE Software, every employee is financially rewarded as if he or she were an owner. "We have a profit bucket," says CEO David Selden, "and once it gets filled, we distribute a percentage to every employee, based on his or her salary. Sometimes we fill more than one bucket in a month, and the distribution generates quite a bit of momentum. It focuses people on what really matters in our business—making a profit."

ProBATE also practices open-book management, in which company financial statements are available to employees. (Of course, private firms aren't legally required to do this.) "They can see the cost of running a business instead of thinking the owners are just taking all the money and stuffing it into their pockets," says Seldon. "They also see where their part does matter."

Many accounting and consulting firms practice versions of employee ownership by creating equity plays for nonpartner employees and speeding up the partnership track. Deloitte Touche Tohmatsu has a so-called *phantom stock plan* known as Career Shares, which lets eligible employees invest in a choice of mutual fund portfolios. Accenture gives eligible employees *e-units,* representing a chunk of the company's venture fund, payable to the employee depending on tenure (with a required minimum of three years) and accomplishments.

On the investment banking side, Goldman Sachs recently made two million shares of stock available to 8,000 junior associates, while J. P. Morgan Chase started a $1 billion program to invest in employees' Internet ideas. The *Washington Post* wrote, "The concept, essentially, is to give star employees a place to exercise their dot-com desires within the security of an established company."

One caveat: Remember that most employee ownership incentives remain in effect only while you remain an employee. There are CEOs who have a reputation for "falling out of love" with an employee just before his options are vested.

Consider Consulting

Another avenue for making significant money in the high-tech arena is independent consulting.

Most consultants specialize in a specific technology or group of technologies, offering their expertise to companies on a per-project or per-day advisory basis. The financial compensation can be significant—up to $300 an hour or more—but there are no paid benefits and no equity potential.

Consulting isn't for everyone. To become a consultant, you'll need significant expertise, impressive credentials, a track record of success with good companies, and a network of contacts at several firms to serve as a source of assignments.

You also need a "consultant's personality." This means being a self-starter and self-motivator, capable of working on your own; being willing and able to continually market your services; and being able to handle the psychological stress of having no fixed income, no benefits, and wildly fluctuating cash flow.

For many consultants, especially those new to the field, determining the proper rate of compensation is the single greatest challenge. Companies usually have a fixed rate structure in place for contractors—except in the software and dot-com worlds, where they seem willing to pay whatever is necessary to get the right person to do the job. Here are some strategies that have worked for one technology consultant.

> When business people call me about a consulting job, they often ask whether I'm currently working. The answer tells them whether or not I'm desperate, which of course has a huge impact on my bargaining stance. Since I've always been working, I've never had to lie—but I would suggest that you say you're working whenever you're asked.
>
> When negotiations begin, I will do anything I can to avoid offering the first compensation figure. I want them to put me in the ballpark. To force this, I will press them with questions such as, "What have you paid for similar work in the past?" or "What amount do you have budgeted for the entire project?"
>
> If I must cite the first number, I will start very high just to test their reaction—for example, 50 percent to 75 percent higher than what I usually get. Sometimes they laugh, sometimes they say, "We can't go that high." Either way, they generally give me a number at that point.
>
> In determining my minimum dollar figure for a consulting job, I take several factors into account. The shorter the assignment, the higher the hourly rate. The cost of travel and other expenses is a factor, and if the assignment will involve several weeks or months at the company's location, so is the local cost of living. (Housing and other costs in Silicon Valley are extremely high.) I may also adjust my

demands based on the long-range potential of the assignment. A job that will provide an opportunity to develop valuable skills, contacts, or experience may be worth taking even if the compensation is slightly low.

Again, not everyone is cut out to be a consultant. But if you have the right temperament for it, it can be a great way of life—and quite lucrative.

Start Your Own Company

The classic American path to wealth, both in the high-tech world and in other industries, is entrepreneurship: Come up with a great product or service and market it.

Of course, starting and running a business is a big undertaking, and entrepreneurship is an enormously complicated field that has inspired hundreds of books. This brief section only provides some of the key considerations that go into successful entrepreneurship, which include the following.

- *Fill a void by meeting an unmet customer need.* It's not enough to come up with a new technology that's "cool" or impressive. You have to build a product or service that serves people by making things easier, saving time, or saving money. The main reason most new companies fail is that they don't stay close to their customers and their needs.
- *Develop a realistic business plan.* Your new company must fit the traditional business parameters by making money and providing a good return on investment. Especially during the early years, maintaining adequate cash flow can be enormously difficult, and failing to do so has been the downfall of countless would-be entrepreneurs. Even a company that sells a popular new product that serves a genuine need can fail if the rate at which revenues flow into the company's coffers isn't sufficient to offset expenditures. It takes financial expertise to anticipate this problem. If you don't have it, partner with someone who does.

- *Keep costs low.* Many businesses fail because their owners spent too freely. It's tempting to invest in handsome office space, an expensive computer system, or a glossy Web site, and it's easy to justify such spending by saying, "We need to look good to impress our customers." Unfortunately, such drains on your company finances can be fatal to the business, and often are. Most successful entrepreneurs are penny-pinchers, especially when it comes to personal perks and benefits.
- *Expect to work harder than ever.* Most company founders agree that their early years in business were the most difficult, demanding times of their lives. Forget about free weekends, evenings, and vacations: When you are the ultimate authority, and the company's success or failure rides on your shoulders, then the cliche of "24/7" becomes literally, exhaustingly true. (And your spouse, children, and friends are likely to find themselves on tap as well.)
- *Build your business around your passion.* Having said all this, it's obvious that the personal costs of entrepreneurship are high—and because 70 percent of all new businesses fail within two years, the odds of success are relatively low. Therefore, if you want to pursue the entrepreneur's path, try to focus on a business idea that incorporates a technology, a market segment, or a business interest that you truly love.

If you do this, you'll at least have some fun mixed in with the long hours and the inevitable frustrations that will come from pursuing your dream.

12

HIT THE GROUND RUNNING

"Don't think you've got it all figured out right from the start. Believe me, most things have been thought of already. Just listen, listen, listen, and listen some more. Eventually you will find an area that is being underserved, and you can capitalize on your ideas then."

—STEVE ADELE, COO, Nutros.com

On the Path to Success

Forty percent of new executives fail within their first eighteen months on the job. Don't let that to happen to you.

You may think the transition to your new job will be easy. Perhaps you're moving to a similar size or similar type of company; perhaps you've known the CEO for years; perhaps the interviews went great, the environment feels friendly, and so on. Nonetheless, almost no job transition is trouble-free. Even if you've made a minor job switch, you can still expect major differences. And if you've made a significant job switch, you can expect major, major differences.

One executive describes his first thirty days in a new job as "traumatic"—and that's after moving from a senior position at one large, old company (IBM) to another (Xerox). Imagine how he'd feel if he'd jumped to a start-up!

It's not only the uninitiated who undergo shock when changing jobs. It's the same for high-ranking executives. One search firm president sends each of his successful CEO candidates to a "boarding seminar." He explains: "We put the new CEOs through exercises based on exhaustive research we've done on making the first thirty to one hundred days effective. And every time we put someone through the program, we refine it for the next one."

You, not the company that you are joining, have to deal with the change. So make a firm decision to adjust to your new employer's and co-workers' styles—they won't be trying to adjust to yours. But remember, it's a matter of adjustment, not compromise.

Go in with a desire to succeed and do great work. Don't just care about pleasing your boss. Care about quality in relating with people. Care about quality in results. Care about quality in every aspect of your work.

Some Hints from the Pros

Your first thirty days at a new job will set the tone for your tenure with the company. To ensure success, here are some helpful guidelines from people who've lived through more than one career transition.

- **Before starting, learn all you can about the job.** If you don't know your duties, responsibilities, and scope of work before getting on board, you'll be in trouble from day one.

Phyllis Rich, vice president of information resources at NEI, speaks about "Bill," a new hire she's been getting good feedback on. Bill took a few days from his between-jobs vacation to come in early and do some homework on the company and industry, which helped him get up to speed right away. Rich says, "I suggested he meet senior managers, and he pushed to get on their calendar. That was good." Bill's success at NEI reflects his gung-ho attitude.

• **Show confidence.** You should know who you are and what you bring to the table. Remember these facts and demonstrate them to others through your behavior. If you don't, some of your new colleagues will run roughshod over you, especially those who had hoped to get your job, hoped others would get it, or for some reason are rooting for your boss to fail. (And don't assume there are no such people; there are!) As Edana Desatnick, who works on worldwide organizational development at Merck, puts it, "If you have self-doubt, they will cut you off in a heartbeat."

• **Make the right first impression.** From the first time you meet your staff, your manner has to win them over to your way of doing things. Be consistent with the "super" person they hired. Reinforce the image you sold in the interview. Prepare fifteen-second answers to the key questions you will get, which will likely include:

- Where did you come from? Why did you leave?
- What have you done?
- What are planning to do here?

• **Study how things work.** Ask and listen. Then shut up and observe. Arrange one-on-one or group meetings at breakfast and lunch, at as many levels as possible. (Don't be offended if people decline your initial invitation. Ask again two weeks later.) Invite people to tell you their top three issues, problems, or items they want assistance on. Meet with customers and ask the same. Always ask as many questions as you possibly can while volunteering helpful information as well.

After a week of this, you should know a lot about the company and the people you are working with. The crucial information includes:

- How do they view your role?
- What can you and your group do to help your peers?
- What should be your agenda for the next ninety days?

- What did they like about how the last person did the job? Where could he or she have done better?
- How does your role fit in with respect to the entire company?

• **Don't be afraid to ask "stupid questions."** Your first two weeks on the job are your one and only chance to ask about the basics *without* appearing stupid. One man says that, because he was too intimidated to ask questions about how something worked, he crashed the company's entire database on his first day on the job. (He probably didn't make a good first impression.)

• **Listen intently even to conversations you are not involved in.** "You can quietly take in tons of information about the work environment, the products, people's attitudes, and your co-workers simply by sipping your coffee in the break room," says software engineer Kevin Werner. He's right.

• **Find out who the informal influencers are.** Steve Cornforth, director of human resources at TrueSpectra, shrewdly observes, "When exterminators come in to rid your house of mice, there is what they call a 'house mouse.' The entire colony takes guidance from that one, and until it's gone the rest won't be. The same thing is true in a new job. You need to find out who the house mouse is and give that person respect, because others take their guidance from him or her."

• **As much as you think you've learned, subtly ask again to verify.** For example, you want to know for certain how your boss likes to communicate and be communicated with. Should you use face-to-face conversations, telephone, voice mail, or e-mail? Does the boss want summaries or full-blown details? How frequently should you make contact? What's too much and what's too little? How does he or she like meetings handled? Answer questions like these by looking, listening, and asking.

• **Don't talk about yourself or your past successes.** You were hired in large measure because of work you've done elsewhere. Because of that, you may have a mistaken tendency to say, "We did it this way at XYZ."

Don't do it. Instead, take what you've learned from your previous employers, add that to what you're learning on the new job, and come up with a new strategy.

• **Never think or act like you know it all.** Never try to impress people with what you know. And don't tell others how to do their job.

• **Focus on relationships.** Become a trusted member of the team—up and down the ladder, inside and outside the company. You'll never recover if you start out as someone unlikable. Build relationships and immerse yourself in them. (But don't overcommit to any one person or align yourself with a particular group too early. Take some time to see who is genuine and who is just talk.)

One Fortune 100 executive confides that in his company, 75 percent of new executives fail within five years. The usual reason is that they didn't focus on building good relationships early on and were never able to recover from that misstep.

• **Build credibility quickly with an early success.** Reinforce why they were right to hire you. "Get into the success zone as quickly as possible," says Christina West, founder and COO of Cater City. Be courteous, polite, and respectful as you decisively take some action. Perhaps there is a small idea the company is struggling with. Take on that one, or invent something else to work on. But do something successful and worthwhile sooner rather than later.

• **Take a lot of initiative without requiring direction.** "Dive right in and get your hands dirty right away," says Eric Weissmann, CEO of Decisioneering. "Don't take the cerebral approach." Come in with, "This is what we should do and here's how I think we should do it." Don't pester the boss with, "Now what should I do?"

Mark Miller of Right Management Consultants explains it well. "Plan to have a plan and be proactive about it. What do you want to accomplish with your new work group or in your new position? What are your opportunities and what are the obstacles to accomplishing them?

Write it all down, make it into a short presentation for your boss, and schedule an appointment with him or her to review your plan. Being proactive in this way allows you to quickly determine what's important and to whom. It also allows you to set the agenda, instead of reacting to everyone else's."

- **Do more than people expect.** Set the right tone during your first weeks on the job by putting in a little more effort than what is asked of you. For example, arrive earlier and leave later than normal working hours.

- **Be exemplary in your attitudes and behavior.** When you make a mistake, acknowledge it, correct it, and don't do it again. Avoid gossip. And remain positive, whatever happens.

- **Stay in touch with the people who brought you on board.** Speak regularly with whoever helped you get into the company—your personal contact, the headhunter, the venture capitalist, the hiring executive. If you're successful in your new job, it reinforces their decision to hire you. If things aren't going well, they might be a source of good advice—or a resource for ways to get out.

EPILOGUE:
IF I KNEW THEN

In my conversations with numerous successful people, I always try to ask, "What do you know now that you wish you'd known at the start of your career?" Here, along with my advice, are some of their answers. My hope is that they'll provide you with some insights into what makes and breaks a career so that you can avoid learning the same lessons the hard and painful way.

"I wish I'd opted to take risks a lot sooner." I have never heard a gray-haired executive say, "I sure wish I hadn't taken so many risks." But I've heard the reverse many, many times.

In a way, it's just common sense. When you're young, you can afford to gamble more, because (1) if you mess up, you'll probably be forgiven ("Aw, he's just a kid," they'll say), and (2) you'll have time later in your career to remedy any serious mistakes you make. Sometimes—especially in times of dramatic, tumultuous change—there's a bigger risk in *avoiding* risk than in taking some.

When young people ask me to explain my experience, I tell them, "In the beginning of my career, I was too ignorant to understand all the risks I was taking. Then as I got older, I got good at taking risks because I had so much practice." Even so, I wish I'd taken even more career risks than I have.

Trust me, you will never regret any well-thought-out gamble you take in your career.

Procrastination and indecision can drain the life from your work. It's better to make a mistake than be indecisive. This applies, for example, to the stay-or-go question considered in Chapter 2. Procrastination from fear or uncertainty often causes people to hang on in untenable job situations too long. "I regret not moving faster," says Steve Cornforth, the director of human resources at TrueSpectra. "I would have left places two years earlier every time. Things were over when I didn't think they were."

(Ironically, those at the other end of an employee's stay-or-go decision often feel the same. When I talk with managers who have terminated poor-performing employees, I usually hear, "It was difficult to do, and I waited too long. I should have done it sooner.")

So when you feel reasonably certain about the right course of action, don't hesitate—dive in and go for it. Hopefully you'll end up like the executive who told me, "Looking back, I have never moved too quickly to address a problem."

"I wish I'd realized the importance of people skills." Every company, whether in the high-tech sector or not, is based on people. No wonder virtually every businessperson I speak to wishes he or she had understood people better and sooner.

One key aspect of interpersonal skills is understanding the political side of business. Some idealistic young workers would like to find a company where there is no political side. I can assure you that if a company has more than one employee, it also has politics. The sooner you understand how to deal with it, the fewer mistakes you'll make.

Dealing with politics doesn't mean learning to fight dirty, to manipulate, to backbite, or to engage in power plays. It does mean learning to recognize the role of ego, emotions, and personality in business decision-making, and learning how to take those factors into account as you strive to accomplish your goals. It means learning how to appeal to other people's deep-seated anxieties, desires, insecurities, dreams, and ambitions in such a way as to transform those potentially negative energies into positive, constructive forces.

This is one area where your network of personal contacts can be invaluable. Every experienced businessperson has encountered politically fraught situations in his or her career, and most people love to recount their war stories. Learn from them how to more effectively deal with this aspect of interpersonal relationships.

"I wish I'd started out with greater self-awareness." Many business people have told me that if they'd had greater self-knowledge, a greater ability to accurately gauge their own strengths and weaknesses, they would have made better decisions. Eric Weissmann, chairman of Decisioneering, speaks for many when he comments, "I would have avoided evaluating opportunities on paper versus knowing what kind of person I am and where would I best fit. I was an engineer in a big company, and I was in trouble every week for being myself. I was quite decisive, commandeering resources, not formal about paperwork, and pushed hard toward a vision. It's what I need as an entrepreneur—but it didn't work in that big company."

You may want to consider aptitude and personality testing to help you recognize your natural psychological tendencies. Some large organizations will provide such testing; if your company doesn't, consider arranging it for yourself. (It's not very expensive.) Many people have found that psychological testing gave them the personal insight they needed to begin shaping their careers along successful, self-reinforcing patterns.

Others have achieved the same kind insight through courses and books on psychology, philosophy, or religion; others through psychotherapy or other counseling; still others from simple introspection like that involved in our compass setting exercise in Chapter 3. Whatever your method, follow the advice of the Greek philosophers—"Know thyself"—and strive to do it sooner rather than later.

"I wish I'd worked harder on networking." Here's what one newly promoted CEO says about this topic: "I wish I'd understood the importance of staying in touch with people. You never know when, why, or how you might need someone's services . . . and I wish I hadn't burned some bridges. I'm finding I have to play catch-up now." If a CEO says this, what about you?

We've already talked about the importance of establishing a strong, active network. As you embark on a new job, whether it's your first in the high-tech world or your tenth, move your networking a step further by building a small adviser/mentor team that touches multiple aspects of the corporate world. Include people in your industry as well as in banking, investing, government relations, law, human resources, and so on.

Wrong as it may seem, the old saw is correct: *Who* you know is as important as *what* you know. The business friendships that you develop, nurture, and continually rekindle are what open doors. It's never too late to work on those friendships, even if you've neglected them for a time. Call and say, "I think of you often, and I wanted to call and see what you are up to." Words like that are welcome to anyone no matter how long it's been since you contacted them last.

As another saying has it, "Schmooze or lose."

"I wish I'd tried to get broader work experience." Many long-term high-tech professionals I spoke to regret the fact that they'd waited for opportunities to be handed to them instead of actively seeking them out.

"I wish I had forced myself into projects and teams that were doing new and exciting things," one long-timer remarked. Another added, "I wish I hadn't let up on learning." Regardless of how many years you've worked, what employers want to see is what you've learned and accomplished *while* working.

Whatever job you hold, don't focus so much on titles, promotions, fringe benefits, or even salary. Instead, focus on lasting assets: learning, experience, change management, and exposure to high-performance people and teams. These are the assets that will form your foundation for future success.

"I wish I'd made career decisions that fit my personal priorities." Too often people in business tend to go with the flow instead of directing their own course. And if the "flow" is pretty good, they keep following it, managing their careers passively rather than actively. Inevitably, one day, the flowing river meets the Hoover Dam, and it's all over. Getting that wake-up call is pretty scary, and it generally comes too late.

You have a right is to have your career expectations met, but they won't be met if you yourself don't know what they are, and if you don't communicate them clearly to the right people.

So begin now to clearly formulate your career and personal objectives and to pursue jobs that match those objectives. Do you want to experience life abroad? Find out what you need to become a manager in an international division of a global company. Are flexible hours and work location of crucial importance to you? Research companies with a reputation for open-ended work arrangements and target them for your next job. Is running your own company your ultimate dream? Get to work on developing the financial, marketing, and administrative skills you'll need to pursue the entrepreneurial path.

"I wish I had realized that I knew more than I thought I did." Many of the high-tech veterans I spoke to made some version of this comment: "I gave people way too much credit for knowing more than me." The idea isn't to discount the wisdom of your colleagues and elders or to ignore your mentors but to appropriately value your own experience and your well-thought-out judgments. Pay attention to your own intuition.

This advice is especially important if you come from an unconventional background that others in high tech may not immediately respect. The truth is that many high-tech companies desperately need the people skills, communication skills, marketing savvy, and managerial smarts that professionals from other industries often bring to the table. If you're from a nontechnical field, don't let the techies intimidate you with their jargon. You have a lot to contribute, and don't forget it.

One information technology manager who has been down this route understands the syndrome all too well. "I allowed myself to feel inadequate because my background wasn't typical to my peers," he now says. "Only later did I see how valuable and rare my skills are. And I stopped apologizing for my lack of technical skills." You can learn from his experience.

"I wish I had developed better negotiation skills." Many professionals, women in particular, bemoan the fact they didn't take firm control of the financial end of their careers sooner. There are many possible explanations for this. In some families, women are raised *not* to talk about money—it's considered unladylike. And many women are raised with a vague feeling that aggressively pursuing their own interests is somehow wrong—it makes them "pushy," "unattractive," "castrating," or "bitchy." If stereotypes like these continue to lurk in the back of your consciousness, do your best to banish them. You have every right to ask for what you deserve, both financially and in every other way—in fact, it's your responsibility to do so.

If you still find it hard to take a stand on your own behalf, think about it this way: Don't you owe it to those who love you (spouse, loved ones, children, parents) to be independent, strong, and self-reliant? If you can't fight for yourself, do it for them!

"I wish I had spent more time managing my career." Fortunately for you, simply by reading this book you've taken a big step along the path of actively managing your career. Thus, you've positioned yourself to avoid one of the most common and painful regrets shared by many high-tech veterans.

As you've learned in this book, it's important to take the time to set your compass, to sketch your own career mission, to evaluate your experience and skills, and to develop a network of professional contacts. And it's crucial to revisit all these areas periodically so as to make steady progress toward your career and life goals.

One CEO told me, "I wish I had been more strategic, more thoughtful about my career. I didn't think about what would happen downstream. Things were good, things went well. But I wish I had had some sort of plan. I went in various directions, often because someone said, "It's a good idea," and so I took it. I should have evaluated what I was doing, what I wanted to do, and the skills that I needed to do it."

"I wish I'd thought bigger." Finally, many of the professionals I spoke to commented, in one way or another, "I wish I had put myself into more challenging, intimidating situations earlier on, both personally and professionally." Put simply: If you accept jobs that don't challenge your capabilities, stretch your talents, engage your energies, and draw you closer toward your ultimate goals, chances are good that you'll *never* reach those goals.

It's easy to understand why some people make this mistake early in their careers. When you're young, you're still trying to figure things out—to see whether or not you can be successful in corporate America. Sometimes it takes many years to realize, "Hey, I really could go somewhere if I work at it." Unfortunately, by the time you make this realization, it may be too late.

My advice: Set your sights higher, *earlier*. Imagine how high you could go; then set a target two or three levels higher than that. Start thinking and acting like you're there *now*. After all, what's the worst thing that can happen? That you'll fall short of those super-ambitious targets, and "settle for" the goals you originally set? How bad would that be?

In Conclusion

To me, it's striking that even the most successful high-tech professionals—vice presidents, CEOs, company chairpersons—have regrets over the opportunities they failed to seize in the early years of their careers. Perhaps taking their advice (and their warnings) seriously can help you to feel better about your own choices. I hope so. In fact, my fondest hope is that one day, when I'm interviewing you for some future book, and I ask, "What would you have done differently?" you'll look me straight in the eye and say with a smile, "I have no regrets."

APPENDIX: SAMPLE EMPLOYMENT CONTRACTS

The following documents provide you with a comparative sample of employment agreements you are apt to be offered. There is, however, no substitute for your own careful consideration to avoid missing a key element in a contract. You will want to take steps to ensure complete legal protection for yourself by having your attorney review your employment agreement.

Sample Executive Employment Agreement

This Executive Employment Agreement ("Agreement") is made and effective this [*date*], by and between [*company*] ("Company") and [*executive*] ("Executive").

NOW, THEREFORE, the parties hereto agree as follows:

1. Employment.

Employment in accordance with the terms of this Agreement and the terms of employment applicable to regular employees of Company. In the event of any conflict or ambiguity between the terms of this Agreement and terms of employment applicable to regular employees, the terms of this Agreement shall control. Election or appointment of Executive to another office or position, regardless of whether such office or position is inferior to Executive's initial office or position, shall not be a breach of this Agreement.

2. Duties of Executive.

The duties of Executive shall include the performance of all of the duties typical of the office held by Executive as described in the bylaws of the Company and such other duties and projects as may be assigned by a superior officer of the

Company, if any, or the board of directors of the Company. Executive shall devote his or her entire productive time, ability, and attention to the business of the Company and shall perform all duties in a professional, ethical, and businesslike manner. Executive will not, during the term of this Agreement, directly or indirectly engage in any other business, either as an employee, employer, consultant, principal, officer, director, advisor, or in any other capacity, either with or without compensation, without the prior written consent of Company. In addition to the duties described herein, Executive is also authorized and directed to do the following: [*list any other specific duties and authorizations*].

3. Compensation.

A. Executive will be paid compensation during this Agreement as follows: A. A base salary of [*salary, e.g. $75,000 (seventy-five thousand dollars)*] per year, payable in installments according to the Company's regular payroll schedule. The base salary shall be adjusted at the end of each year of employment at the discretion of the board of directors.

B. An incentive salary equal to [*list incentive pay rate*] of the adjusted net profits (hereinafter defined) of the Company beginning with the Company's year-end [*list first year of incentive salary*] and each fiscal year thereafter during the term of this Agreement. "Adjusted net profit" shall be the net profit of the Company before federal and state income taxes, determined in accordance with generally accepted accounting practices by the Company's independent accounting firm and adjusted to exclude: (i) any incentive salary payments paid pursuant to this Agreement; (ii) any contributions to pension and/or profit sharing plans; (iii) any extraordinary gains or losses (including, but not limited to, gains or losses on disposition of assets); (iv) any refund or deficiency of federal and state income taxes paid in a prior year; and (v) any provision for federal or state income taxes made in prior years which is subsequently determined to be unnecessary. The determination of the adjusted net profits made by the independent accounting firm employed by the Company shall be final and binding upon Executive and Company. The incentive salary payment shall be made within thirty (30) days after the Company's independent accounting firm has concluded its audit. If the final audit is not prepared within ninety (90) days after the end of the fiscal year, then Company shall make a preliminary payment equal to fifty percent (50%) of the amount due based upon the adjusted net profits preliminary determined by the independent accounting firm, subject to payment of the balance, if any, promptly following completion of the audit by the Company's independent accounting firm.

The maximum incentive salary payable for any one year shall not exceed [*maximum incentive compared to base*] of the then applicable base salary of Executive.

4. Benefits.

A. Holidays. Executive will be entitled to at least [*number of paid holidays*] paid holidays each calendar year and [*number of personal days*] personal days. Company will notify Executive on or about the beginning of each calendar year with respect to the holiday schedule for the coming year. Personal holidays, if any, will be scheduled in advance subject to requirements of Company. Such holidays must be taken during the calendar year and cannot be carried forward into the next year. Executive is not entitled to any personal holidays during the first six months of employment.

B. Vacation. Following the first six months of employment, Executive shall be entitled to [*number of vacation days*] paid vacation each year.

C. Sick Leave. Executive shall be entitled to sick leave and emergency leave according to the regular policies and procedures of Company. Additional sick leave or emergency leave over and above paid leave provided by the Company, if any, shall be unpaid and shall be granted at the discretion of the board of directors.

D. Medical and Group Life Insurance. Company agrees to include Executive in the group medical and hospital plan of Company and provide group life insurance for Executive at no charge to Executive in the amount of [*amount of group life insurance*] during this Agreement. Executive shall be responsible for payment of any federal or state income tax imposed upon these benefits.

E. Pension and Profit Sharing Plans. Executive shall be entitled to participate in any pension or profit sharing plan or other type of plan adopted by Company for the benefit of its officers and/or regular employees.

F. Automobile. Company will provide to Executive the use of an automobile of Executive's choice at a gross purchase price not to exceed [*maximum purchase price*]. Company agrees to replace the automobile with a new one at Executive's request no more often than once every two years. Company will pay all automobile operating expenses incurred by Executive in the performance of an Executive's company duties. Company will procure and maintain in force an automobile liability policy for the automobile with coverage, including Executive, in the minimum amount of $1,000,000 combined single limit on bodily injury and property damage.

G. Expense Reimbursement. Executive shall be entitled to reimbursement for all reasonable expenses, including travel and entertainment, incurred by Executive in the performance of Executive's duties. Executive will maintain records and written receipt as required by the Company policy and reasonably requested by the board of directors to substantiate such expenses.

5. Term and Termination.

A. The Initial Term of this Agreement shall commence on [*start date*] and it shall continue in effect for a period of [*initial term*]. Thereafter, the Agreement shall be renewed upon the mutual agreement of Executive and Company. This Agreement and Executive's employment may be terminated at Company's discretion during the Initial Term, provided that Company shall pay to Executive an amount equal to payment at Executive's base salary rate for the remaining period of Initial Term, plus an amount equal to [*sevarance pay during initial term*] of Executive's base salary. In the event of such termination, Executive shall not be entitled to any incentive salary payment or any other compensation then in effect, prorated or otherwise.

B. This Agreement and Executive's employment may be terminated by Company at its discretion at any time after the Initial Term, provided that in such case, Executive shall be paid [*severance pay after initial term*] of Executive's then applicable base salary. In the event of such a discretionary termination, Executive shall not be entitled to receive any incentive salary payment or any other compensation then in effect, prorated or otherwise.

C. This Agreement may be terminated by Executive at Executive's discretion by providing at least thirty (30) days prior written notice to Company. In the event of termination by Executive pursuant to this subsection, Company may immediately relieve Executive of all duties and immediately terminate this Agreement, provided that Company shall pay Executive at the then applicable base salary rate to the termination date included in Executive's original termination notice.

D. In the event that Executive is in breach of any material obligation owed Company in this Agreement, habitually neglects the duties to be performed under this Agreement, engages in any conduct which is dishonest, damages the reputation or standing of the Company, or is convicted of any criminal act or engages in any act of moral turpitude, then Company may terminate this Agreement upon five (5) days notice to Executive. In event of termination of the agreement pursuant to this subsection, Executive shall be paid only at the then applicable base salary rate up to and including the date of termination. Executive shall not be paid any incentive salary payments or other compensation, prorated or otherwise.

E. In the event Company is acquired, or is the nonsurviving party in a merger, or sells all or substantially all of its assets, this Agreement shall not be terminated and Company agrees to use its best efforts to ensure that the transferee or surviving company is bound by the provisions of this Agreement.

6. Notices.

Any notice required by this Agreement or given in connection with it, shall be in writing and shall be given to the appropriate party by personal delivery or by certified mail, postage prepaid, or recognized overnight delivery services;

If to Company:

[*list company*]

[*list company address*]

If to Executive:

[*list executive*]

[*list executive address*]

7. Final Agreement.

This Agreement terminates and supersedes all prior understandings or agreements on the subject matter hereof. This Agreement may be modified only be a further writing that is duly executed by both parties.

8. Governing Law.

This Agreement shall be construed and enforced in accordance with the laws of the state of [*list state of governing law*].

9. Headings.

Headings used in this Agreement are provided for convenience only and shall not be used to construe meaning or intent.

10. No Assignment.

Neither this Agreement nor any or interest in this Agreement may be assigned by Executive without the prior express written approval of Company, which may be withheld by Company at Company's absolute discretion.

11. Severability.

If any term of this Agreement is held by a court of competent jurisdiction to be invalid or unenforceable, then this Agreement, including all of the remaining terms, will remain in full force and effect as if such invalid or unenforceable term had never been included.

12. Arbitration.

The parties agree that they will use their best efforts to amicably resolve any dispute arising out of or relating to this Agreement. Any controversy, claim or dispute that cannot be so resolved shall be settled by final binding arbitration in accordance with the rules of the American Arbitration Association and judgment upon the award rendered by the arbitrator or arbitrators may be entered in any court having jurisdiction thereof. Any such arbitration shall be conducted in [*location of arbitration*], or such other place as may be mutually agreed upon by the parties. Within fifteen (15) days after the commencement of the arbitration, each party shall select one person to act arbitrator, and the two arbitrators so selected shall select a third arbitrator within ten (10) days of their appointment. Each party shall bear its own costs and expenses and an equal share of the arbitrator's expenses and administrative fees of arbitration.

IN WITNESS WHEREOF, the parties hereto have executed this Agreement as of the date first above written.

[*company signature block*] [*executive signature block*]

[*company signature block*] [*executive signature block*]

[*company signature block*] [*executive signature block*]

[*company signature block*] [*executive signature block*]

Source: www.Lawsmart.com

Employment Agreement

THIS AGREEMENT is made, effective the _______ day of __________, by and between __________________, a corporation duly organized and existing under the laws of the State of _____________, with a place of business at _______________________________________, hereinafter referred to as "Employer", and ____________________________ whose address is ____________________________, hereinafter referred to as "Employee."

SECTION ONE: EMPLOYMENT

A. Employer hereby employs, engages, and hires Employee and Employee hereby accepts and agrees to such hiring, engagement, and employment, subject to the general supervision and pursuant to the orders, advice, and direction of Employer.

B. Employee shall perform such duties as are customarily performed by one holding a position in other, same, or similar businesses or enterprises as that engaged in by Employer, and shall also additionally render such other and unrelated services and duties as may be assigned to Employee from time to time by Employer.

SECTION TWO: BEST EFFORTS OF EMPLOYEE

Employee will at all times faithfully, industriously, and to the best of Employee's ability, experience, and talents, perform all of the duties that may be required of and from Employee pursuant to the express and implicit terms of this agreement, to the reasonable satisfaction of Employer.

SECTION THREE: TERM OF EMPLOYMENT

This is an employment agreement "at will" under ______________ [*name of state*] law. Either Employee or Employer may terminate this agreement at any time, for any reason with or without cause. Termination of this agreement shall be complete upon the terminating party tendering to the other party written notice of his or her intent to terminate this agreement.

SECTION FOUR: COMPENSATION OF EMPLOYEE

A. Employer shall pay Employee, and Employee shall accept from Employer, in full payment for Employee's services under this agreement, compensation at the rate of __________ Dollars ($__________) per __________, payable twice a month on the _____________ and _____________ days of each month while this agreement shall be in force.

B. Employer shall reimburse Employee for all necessary expenses incurred by Employee while traveling pursuant to Employer's directions.

C. Employee may, in the sole and absolute discretion of the Employer, from time to time receive increases or bonuses in his or her pay.

D. In addition to the compensation Employee receives hereunder, Employee is eligible for participation in the fringe benefit programs established by Employer on fulfillment of the eligibility requirements for each program. Employer may, without notice, modify or discontinue any fringe benefit program that it maintains.

E. Employee is entitled to a paid vacation of __________ days per year.

SECTION FIVE: OTHER EMPLOYMENT

Employee shall devote all of Employee's time, attention, knowledge, and skills solely to the business and interest of Employer, and Employer shall be entitled to all of the benefits, profits, or other issues arising from or incident to all work, services, and advice of Employee, and Employee shall not, during the term of this agreement, be interested directly or indirectly, in any manner, as partner, officer, director, shareholder, advisor, Employee, or in any other capacity in any other business similar to Employer's business or any allied trade; provided, however, that nothing contained in this section shall be deemed to prevent or to limit the right of Employee to invest any of Employee's money in the capital stock or other securities of any corporation whose stock or securities are publicly owned or are regularly traded on any public exchange, nor shall anything contained in this section be deemed to prevent Employee from investing or limit Employee's right to invest Employee's money in real estate.

SECTION SIX: RECOMMENDATIONS FOR IMPROVING OPERATIONS

Employee shall make available to Employer all information of which Employee shall have any knowledge and shall make all suggestions and recommendations that will be of mutual benefit to Employer and Employee.

SECTION SEVEN: TRADE SECRETS/PROPERTY RIGHTS

A. Employee shall not at any time or in any manner, either directly or indirectly, divulge, disclose or communicate to any person, firm, corporation, or other entity in any manner whatsoever any information concerning any matters affecting or relating to the business of Employer, including without limitation, any of its customers, the prices it obtains or has obtained from the sale of, or at which it sells or has sold, its products, or any other information concerning the business of Employer, its manner of operation, its plans, processes, or other data ("Confidential Information") without regard to whether all of the above-stated matters will be deemed confidential, material, or important by others, Employer and Employee specifically and expressly stipulating that as between them, such matters are important, material, and confidential and gravely affect the effective and successful conduct of the business of Employer, and Employer's good will, and that any breach of the terms of this section shall be a material breach of this agreement.

B. During the course of his or her employment with Employer, Employee may be producing software code and other items related to computers and/or their peripherals. All inventions, designs, developments, formulas, patterns, devices, compilations of information, records and specifications, computer programs, hardware, software code and/or marketing programs (and any portions thereof) produced and/or conceived by the Employee while employed with Employer and/or that were conceived from the use of equipment, facilities, or other resources of the Employer or which the Employer possessed at the time of the execution of this agreement (all of the foregoing shall be collectively referred to as "Intellectual Property"), shall remain the sole and exclusive property of the Employer. Intellectual Property shall also include any inventions, designs, developments, formulas, patterns, devices, compilations of information, records and specifications, computer programs, hardware, software code and/or marketing programs produced by Employee after the termination of the Employee-

Employer relationship to the extent such items relate in any fashion to an idea, concept or program which was originally conceived or produced while Employee was employed with Employer. The Employee shall promptly disclose and fully inform to the Employer the details of all such Intellectual Property as soon as the same becomes known to Employee. For purposes of this agreement the terms "conceived" and "produced" shall be given the broadest possible interpretation and shall include any thought process during which an idea is created regardless of whether the idea as originally conceived or produced requires alteration to become practical or useful.

C. Employee agrees that he or she has no past, present or future claim or right to ownership of any of the Intellectual Property, any current or future proceeds from the sale of any Intellectual Property or profits derived from the Intellectual Property by the Employer or Employee, and/or any Intellectual Property currently belonging to the Employer or Intellectual Property which is conceived of or produced by the Employee and which becomes property of the Employer pursuant to the terms hereof. To the extent that the Employee may in the future attempt to claim any ownership interest in or legal right to the Intellectual Property or any portion thereof, any current or future proceeds from the sale of the Intellectual Property or the use thereof, and/or any Intellectual Property currently belonging to the Employer or Intellectual Property which becomes property of the Employer hereunder, Employee hereby expressly waives such claims regardless of whether such claims are now known or of an unknown origin and nature.

SECTION EIGHT: TRADE SECRETS AFTER TERMINATION OF EMPLOYMENT

All of the terms of Section Seven of this agreement shall remain in full force and effect for the period of thirty (30) years after the termination of Employee's employment for any reason, and during such time period.

SECTION NINE: COVENANTS OF EMPLOYEE/ NON-COMPETITION AGREEMENT)

A. Employee shall not, directly or indirectly at any time during his or her employment with Employer, and for a period of two (2) years after termination of the Employee-Employer relationship:

1. Solicit or attempt to solicit any employee, agent or contractor of Employer to leave the employment of Employer; or

2. Assist or attempt to assist any person, firm or corporation in any way to solicit any employee, agent or contractor of Employer to leave the employment of Employer.

B. At such time as the employment relationship between Employee and Employer has terminated, Employee shall:

1. Promptly return to Employer, or at Employer's option, destroy all Confidential Information and any documents related to the Intellectual Property, including all copies of documents, notes or materials made by Employee or at her direction; and

2. Certify in writing to Employer that he or she has so complied; and

3. Not use Confidential Information or Intellectual Property or transact business in a manner in any way based upon or utilizing Confidential Information or Intellectual Property.

C. Commencing on the effective date of this Agreement and continuing for a period of two (2) years after termination of Employee's employment with Employer, Employee, either individually or in partnership or jointly or in conjunction with any person or persons, firm, association, syndicate, company or Employer, as principal, agent, shareholder, or in any other manner whatsoever shall not solicit, divert or take away, or attempt to solicit, divert or take away, any customers of Employer, or call upon communicate, advise or consult with, write or respond to, or inform any customer, client or account of Employer, for the purpose of soliciting, selling or recommending conflicting products and service (whether or not such customer, client or account was a customer, client or account previously contacted by Employee while employed by Employer), in any area in which Employer (or its agents or representatives) sells its products and services not to exceed two hundred miles (200) from the outer limits of Franklin County in any direction. This covenant not to compete shall be effective regardless of the reason why Employee was terminated even if such termination was arbitrary, capricious or wrongful.

D. In the event that any or all of the restrictive covenants shall be determined by a court of competent jurisdiction to be unenforceable by reason of their geographic or temporal restrictions being too great, or by reason that the range of activities covered are too great, or for any other reason, they should be interpreted to extend over the maximum geographic area, period of time, range of activities or other restrictions as to which they may be enforceable.

E. The provisions of this Section Nine shall survive the termination of this Agreement by thirty (30) years.

SECTION TEN: EMPLOYEE'S INABILITY TO CONTRACT FOR EMPLOYER

In spite of anything contained in this agreement to the contrary, Employee shall not have the right to make any contracts or commitments for or on behalf of Employer without first obtaining the express written consent of Employer and shall have no authority to write checks on behalf of the Employer. In the event the Employee enters into any agreement or writes any check on behalf of employer which is not specifically authorized in writing by Employer, Employee shall indemnify Employer for any costs incurred by Employer in connection with such agreement or check.

SECTION ELEVEN: MISCELLANEOUS

A. This agreement contains the complete agreement concerning the employment arrangement between the parties and shall, as of the effective date hereof, supersede all other agreements between the parties. Neither party has made any representation with respect to the subject matter of this agreement not specifically included in this agreement nor has either party relied on any such representation in entering into this agreement.

B. This agreement may only be modified by a writing signed by both parties.

C. The invalidity of any portion of this agreement will not and shall not be deemed to affect the validity of any other provision. In the event that any provision of this agreement is held to be invalid, the parties agree that the remaining provisions shall be deemed to be in full force and effect as if they had been executed by both parties subsequent to the expungement of the invalid provision.

D. This agreement shall be interpreted in accordance with Ohio law.

E. The failure of either party to this agreement to insist upon the performance of any of the terms and conditions of this agreement, or the waiver of any breach of any of the terms and conditions of this agreement, shall not be construed as thereafter waiving any such terms and conditions, but the same shall continue and remain in full force and effect as if no such forbearance or waiver had occurred.

F. The titles to the paragraphs of this agreement are solely for the convenience of the parties and shall not be used to explain, modify, simplify, or aid in the interpretation of the provisions of this agreement.

IN WITNESS WHEREOF, each party to this agreement has caused it to be executed at ________________ on the date indicated below.

EMPLOYER:

____

EMPLOYEE:

____

DATE:

____

Source: www.net.set.com

SPECIAL THANKS

A special thanks to the people who contributed their experiences to this book, including:

Steve Adele, Chief Operating Officer, Nutros.com

Steve Aldrich, President, QuickenInsurance

Jay Atlas, Chief Executive Officer, Ariel Corp.

Janice Aune, Chief Executive Officer, Onvoy

Pam Stegora Axberg, Senior Vice President, Central Office, Qwest

John Borden, Chief Operating Officer, LearningSoft

Ron Brown, Chief Executive Officer, Maximation

Ross Brown, Vice President Human Resources, Analog Devices

Alan Button, Vice President, Resonate

Todd Canedy, Chief Executive Officer, AXOXS

Christine Carey, Vice President Web Services, AMI Visions

Katherine Cizynski, Senior Partner, Wiser Partners

Jonathan Cohen, Chief Executive Officer, 20 Tons

Steve Cornforth, Director Human Resources, TrueSpectra

Jeff Cunningham, Partner, Internet Finance Partners

Jerome Davis, President, Business Process Management at Electronic Data Systems

Chris Day, Co-president, Packtion

Charles Dicket, President, e-catalyst

Rod Dobson, President, Employee Services Group, ADP Canada

Dale Douglas, electrical engineer

Michael Doyle, Senior Brand Director, Immedient

Amy Duclos, Chief Executive Officer, Colorado Corporate Search

Phil Duong, Chief Technology Officer, Iconixx

Kristie Eslick, Vice President, International Solutions & Technical Services

Jack Falvey, Founder, MakingTheNumbers.com

Larry Fehd, President, Human Performance Strategies

Terry Garlow, Strategic Alliance Manager, Hewlett-Packard

Mickey Gart, Chief Executive Officer, Gart Development Group

Michael Gellman, Chief Executive Officer, Spiremedia

Jerry Golley, Chief Executive Officer, AMI Visions

Norbert Gottenberg, Recruiting Partner, Kleiner Perkins

Allan Grafman, President, Geyser Technologies

Laura Gross, Attorney, Kaufman & Canoles

Dan Hancock, Gorham Savings Bank

Tom Hardy, Partner, Heidrick & Struggles

Dave Hardie, Partner, Heidrick & Struggles

Ken Hayashi, Certified Public Accountant

Doug Heard, Human Resources Director, Raytheon

Kathleen Ho O'Donnell, Principal, Advent International Corp.

Jennifer Hopkins, Vice President and General Manager, GSU/Agilent Technologies

Dennis Hoppe, President, Hoppe Management Concepts

Bob Hornung, Founder, Onlink.com

Dr. Eileen Howell, Regional Manager, Systems & Computer Technology Corp.

Ernie Howell, Retired President, WPM Systems

Lyle Hurst, Director HP World e-Inclusion, Hewlett Packard

Jeff Irby, Managing Director, KPMG Consulting

Jeff Jacoby, Director of Information Technology, Raytheon

Hal Johnson, Senior Partner, TMP Worldwide

Keith Johnson, Vice President, ThinkArena, FastIdeas Accelerator

Ragesh Jolly, Vice President Marketing, Budget-Rent-A-Car

Sue Kanrich, Training Consultant, F-O-R-T-U-N-E

Tom Keller, Chief Executive Officer, Net Infrastructure

Mark Kimmel, retired venture capitalist and science fiction writer

Kevin King, President and Chief Technologist, Precision Solutions, Inc.

David Klotz, Founder, COMOCEAN

Justin Knecht, Manager, Internet Technologies

Diane Kunkel, Director of Programming, National Technical University

Kent Lassman, Director of Technology and Communication Policy, Citizens for a Sound Economy

Karen Lewis, Raytheon

Jim McBride, Chief Executive Officer, ATMO

Brian McCune, Managing Partner, e-merging technologies group

Jim McGovern, Partner, Zissis McGovern

Lisa Mandell, Vice President Marketing, siteROCK

Bill Manning, Electrical Engineer, NASA

Robert Mayer, Director Human Resources, Packtion

Candace Meyer, Area Director of Sales, Atria

Steve Milovich, Founder, Milovich Partners

Mark Miller, Executive Vice President, Right Management Consultants

Michelle Monfor, Director/Executive Recruiter, TRRG Executive Search

Matthew Moran, President, CaterCity

Herb Morreale, Founder, XOR, Inc.

Chris Nazarenus, President, e-catalyst

Mike Nieset, Vice President, Christina & Timbers

Rick O'Donnell, Director, Office of Policy & Initiatives, Colorado Governor's Office

Gloree Parker-Roden, Senior Vice President Enterprise Services, Pearson Technology Centre

Duane Pearsall, Founder, Columbine Venture Fund

Mary Petryszyn, Director of Engineering, Raytheon Company Rocky Mountain region

Geoff Pike, Managing Director, F-O-R-T-U-N-E

Claude Pineault, Chief Operating Officer, Quebec Software Promotion Centre

Dan Purdy, Demand/Capacity Planning, Agilent Technologies

Sasa Radulovic, President, TrueSpectra

Phyllis Rich, Vice President Information Resources, NEI

Darren Romano, Managing Director, Korn/Ferry

Steven Schimmel, software engineer

Andy Schlossberg, Lead Flash Animator, romp.com,

Markus Schweig, Manager, Microsoft Corporation

Steve Seidman, Chairman, Department of Computer Sciences, Colorado State University

David Selden, Chief Executive Officer, ProBATE Software

Jack Shields, General Partner, Best Capital Ventures

Norman Singer, President, Israel Technology Partners

Patrick Tam, Facility & Regulatory Compliance Manager, Virco

Fred Taraschi, Managing Director, Ken Clark International

Dale Telford, Chief Executive Officer, bITssol Ltd

Karen Thomas, Director Human Resources, Raytheon Strategic Systems Business Unit

Chris Vargas, President, F-Secure

Charles E. Wallace, Managing Partner, Heidrick & Struggles

Eric Weissmann, Chairman, Decisioneering

Kevin Werner, software engineer

Christina West, Chief Operating Officer, CaterCity

David Zodikoff, Director Information Technology, Dell Computer

Index